UPGRADING YOUR AIRPLANE'S AVIONICS

by Timothy R.V. Foster

MODERN AVIATION SERIES

TAB BOOKS Inc.

BLUE RIDGE SUMMIT, PA. 17214

FIRST EDITION

FIRST PRINTING

FEBRUARY 1981

Copyright © 1981 by TAB BOOKS Inc.

Printed in the United States of America

Library of Congress Cataloging in Publication Data

Foster, Timothy R V
    Upgrading your airplane's avionics.

    Includes index.
    1. Airplanes—Electronic equipment. I. Title.
TL693.F66   629.135′028′8   80-28772
ISBN 0-8306-9620-2
ISBN 0-8306-2301-9 (pbk.)

Cover photo courtesy of King Radio Corp.

Dedication and Acknowledgements

For Rex Nicholls, who taught me to fly

The following have been very helpful in the creation of this book: Marty Balk, John Ferrara, Mary Foster, Ken Ross and the people of the avionics manufacturers.

Thank you.

Preface

When I bought my Comanche, it came equipped with 68 pounds of radios and other electronic gear, and it was a pretty good airplane. Today (several thousand dollars later), my Comanche carries 37.4 pounds' worth of avionics, and I think it's a *great* airplane.

Now, it may seem that I've spent an inordinate amount of money to remove 30 pounds from my airplane, but you might be interested in knowing that at least one major U.S. airline is stripping the paint from all of its jetliners. (They figure the move will save them *several hundred thousand* dollars each year in reduced fuel costs.) However, I didn't go to all the trouble and expense of upgrading my avionics just for the minute benefits in aircraft performance or economy.

No, I did it for a number of other reasons—reasons like safety, reliability, versatility, convenience, compliance with FAA regulations, and, yes maybe even just a slight hint of "snob appeal." But foremost, of course, is *safety*. Even though hundreds of airplanes fly every day with *no* avionics at all (and quite legally, too), few things can inspire as much confidence in the air as a well-rounded complement of the latest electronic marvels for communication, navigation, indentification, and control. (And, of course, the skill and knowledge of how to correctly use them.) I wouldn't say that flying a private aircraft is a particularly dangerous pursuit; I *would* say, though, that today's avionics can make the activity just that much more safe.

After all, you wouldn't hop into your car for a Sunday drive with a Victrola sitting on your lap, would you? No? Then why not take a look at your airplane's avionics—or lack of them—and consider updating them?

I think you'll be glad you did.

Timothy R. V. Foster

Contents

Chapter 1
Why Upgrade?

Flying has become a much more complex affair in the last few years, and it is the task of avionics to simplify our lives as much as possible (Fig. 1-1). Many say that the complexities have been brought about by requirements for *more* avionics—thou shalt not fly *here* without an encoding altimeter, thou shalt not fly *there* without DME, thou shalt not fly *here* except IFR, thou shalt not land *there* but that thou holdeth at least a Private License, and canst speak with the tower on 133.05.

An airplane could be said to be a means of transporting a multiplicity of electronic devices between a variety of communication, navigation and interrogation centers for the purpose of providing employment to a large number of overworked civil servants! Another definition I heard once was that an airplane was a device used to transport a pilot and his radios between avionics repair shops! Many frustrated owners have often found this to be the case, but fortunately avionics reliability seems to be improving slightly.

THE GOOD OLD DAYS WEREN'T THAT GOOD

In the late 60's, a new transponder cost at least $1,875 and weighed at least 7.5 pounds. In the early 80's, with all the heavy inflation we have seen, a new transponder costs about $650 and weighs about 3 pounds. And it works better. A typical 1967 "one and a half" NAVCOMM offered 90 COMM channels, cost about $1,200 and weighed 11.3 pounds. You can buy a similar quality unit

Fig. 1-1. Definitely not a candidate for upgrading is this 1980 Cessna Turbo Centurion with a "full house," including radar, flight director, RMI, and autopilot. Avionics by ARC Division of Cessna (courtesy Cessna).

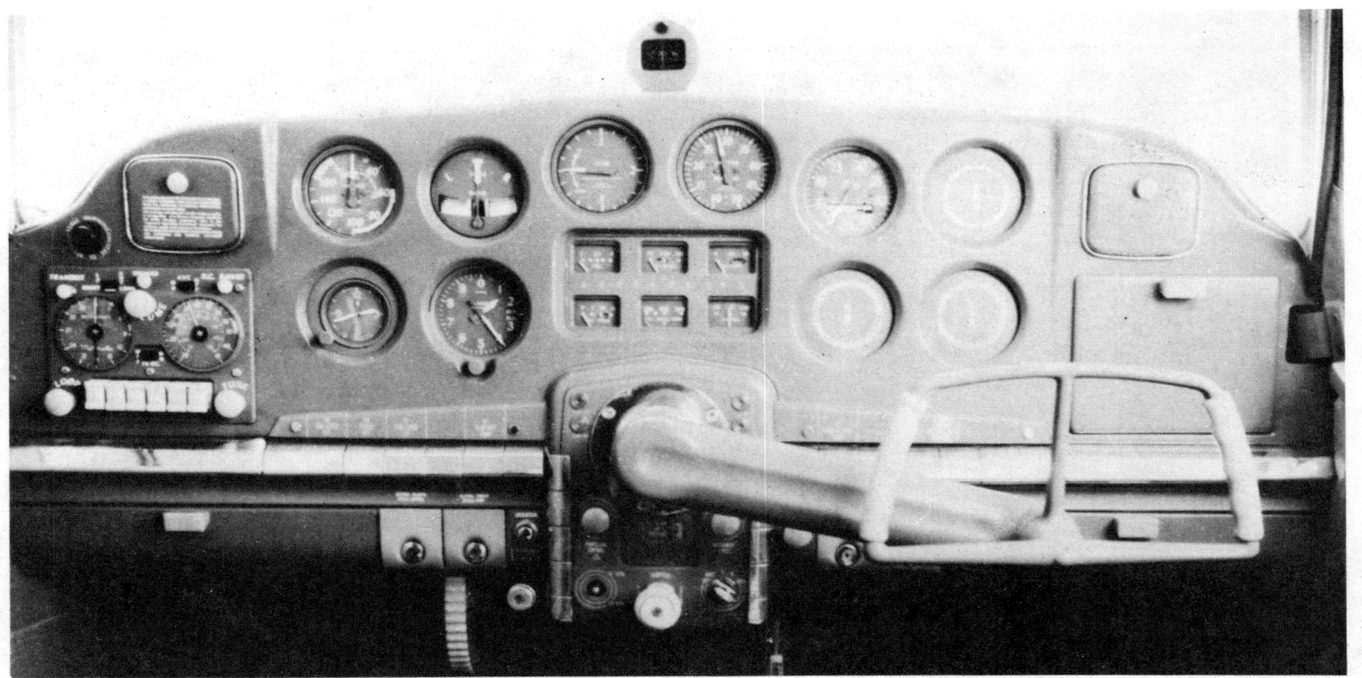

Fig. 1-2. You've come a long way, baby. This is the panel of a 1947 Beechraft Bonanza (courtesy Beech).

Fig 1-3. Here is its current counterpart. Avionics by Collins (courtesy Beech).

today for $2,440, weighing 8 pounds and offering 720 COMM channels.

Weather radar was not available on a single engine aircraft in 1967. To put one in your twin then would have cost you upwards of $5,700, plus installation and radome, and would have increased your gross weight by over 26 pounds. Now you can have radar in certain singles for $5,500 and 13 pounds (plus installation). And you can even get color radar, with your checklists and routes available (for many more dollars).

PILOTS ARE MORE SOPHISTICATED TODAY

The avionics industry has come a long way in those few years (Fig. 1-2, 1-3), which is a good thing, because so has the typical

Fig. 1-4. Why you should upgrade. Nobody should have to settle for this kind of radio today (courtesy Cessna).

pilot. In 1967, there were 429,264 fixed-wing pilots holding at least a private certificate, with 122,573 of these (28.5 per cent) holding an instrument rating. In 1978, there were 566,300 pilots (a 32 per cent increase over 1967) and 226,300 of these (40 per cent) held an instrument rating (an increase of 85 per cent in instrument-rated pilots).

MOST AIRCRAFT FLYING ARE NOT THAT NEW

The majority of general aviation aircraft flying today are not brand new—in fact to a large extent they are quite old. This often means that they have old avionics, and thereby hangs the purpose of this book. It is well known that an airplane never really gets *too* old. You just keep giving it good maintenance, lots of TLC, and a fresh annual inspection each year. Rebuild or replace the engine every 1,500 to 2,000 hours, and it should keep going indefinitely.

MOST AVIONICS ARE NOT THAT NEW

All the while your avionics are aging too, with no similar maintenance schedule (Fig. 1-4). Rubber deteriorates. Wires get brittle and corroded. Tubes lose effectiveness due to heat and many on-off cycles. Occasional voltage spikes may have done small bits of harm to transistors on a cumulative basis, and every so often they all gang up on you and you get a radio failure. Murphy's Law, of course, says that this will take place at the worst possible time.

THE UPGRADING SOLUTION

The solution is to upgrade your airplane's avionics systems as the years go by. The question is how to go about it. The answer is in this book. Where do you start? What do you keep and what do you junk? Can you trade a radio in? Should you buy a used radio? What's the best new radio for the job? That's what this book is all about.

Chapter 2
More Why Upgrade?

There are several reasons a pilot would want to upgrade his or her avionics package. An older airplane may have older avionics that don't work too well, if at all. Or the type of flying the pilot is doing dictates the need for equipment that is not installed, such as a transponder, encoding altimeter or DME. Or the radio frequencies may not be adequate for IFR. Or the pilot simply may not like the brand of avionics installed. Or the pilot may want to put in area navigation (RNAV) only to find that the VOR and DME now installed won't work with the RNAV. Or the ancient appearance of the existing avionics offends the aesthetic eye of the pilot. Or the avionics in the aircraft may not be TSO'd, and the pilot wants them to be. Or the weight of the older avionics exacts a penalty on the airplane that could be removed by going to newer, lightweight equipment.

TECHNICAL STANDARD ORDER (TSO)

TSO means Technical Standard Order, and it refers to a set of specifications about environmental and performance capabilities for each type of radio equipment issued by the FFA. The areas are:

- ☐ Temperature and altitude
- ☐ Humidity
- ☐ Vibration
- ☐ Audio-frequency susceptibility

□ Radio-frequency susceptibility
□ Spurious energy
□ Explosion
□ Electrical performance

To comply with the TSO, a sample of the equipment must undergo certain tests to see if it will operate within certain limits. If it passes these tests, it can be given a TSO-compliance nameplate, which is attached to the equipment (Fig. 2-1).

If a component has been TSO'd, it means that it should meet the limits set by the FAA for compliance. It doesn't necessarily mean that non-TSO'd equipment is worse than TSO'd. It does mean that if it's TSO'd, it meets certain environmental and performance limits.

Older transponders did not have to be TSO'd. All transponders installed in aircraft must now be TSO'd, so an older, non-TSO'd transponder is illegal and may not be used—certainly a reason for upgrading.

Older COMMs had 90 VHF channels—118.0 to 126.9, in 100 kHz steps (118.0, 118.1, 118.2, etc.). Later, they got 360 channels, 118..00 to 135.95 in 50 kHz steps (118.00, 118.05, 118.10, etc.). Then came 720 channels, 118.000 to 135.975 in 25 kHz steps (118.000, 118.025, 118.050, 118.075, etc.). Many older aircraft still have COMMs with 90 or even fewer channels, and these are just no good for IFR flying. You need at least 360 channels for IFR, and one day, you'll need 720. Some old VHF transmitters don't meet the Federal Communication Commission's (FCC's) standards of .005% frequency tolerance, required in all aircraft since 1970 (.003% in all transmitters manufactured after January 1, 1974).

AVIONICS EQUIPMENT REQUIRED FOR FLIGHT

Federal Aviation Regulations (FARs) specify the minimum equipment allowable for flight under visual flight rules (VFR) by day and by night and under instrument flight rules (IRF). These requirements are spelled out in FAR 91.33, as follows:

Day VFR

□ Airspeed indicator
□ Altimeter
□ Magnetic direction indicator
□ Tachometer for each engine

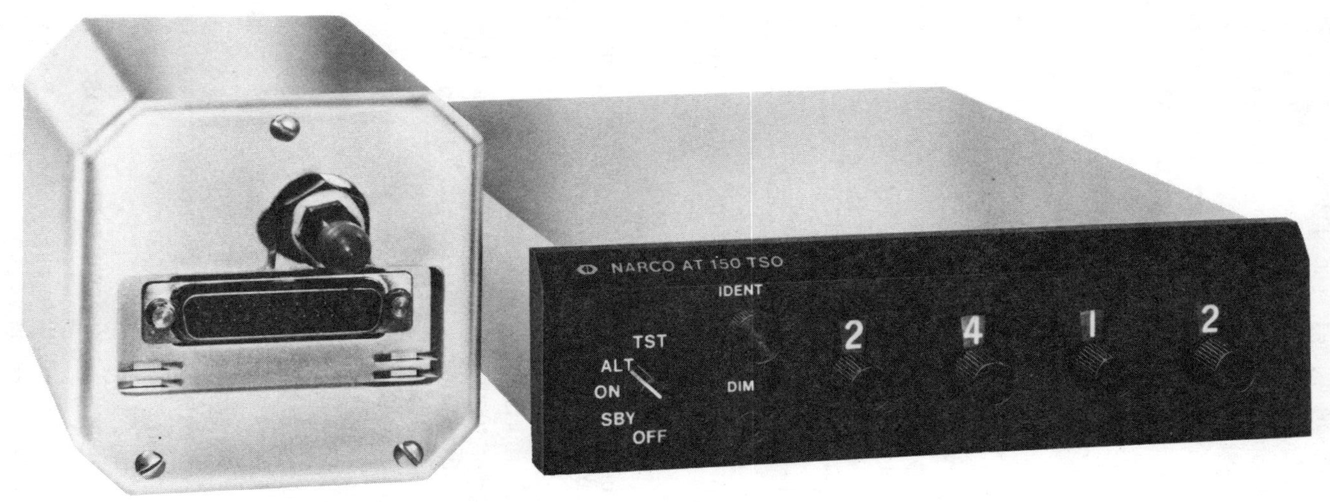

Fig. 2-1. Narco AT 150 TSO'd transponder and AR 500 remote encoder (courtesy Narco).

☐ Oil pressure gauge for each engine
☐ Temperature gauge for each liquid-cooled engine
☐ Oil temperature gauge for each air-cooled engine
☐ Manifold pressure gauge
☐ Fuel gauge indicating the fuel quantity in each tank
☐ Landing gear position indicator if the aircraft has retractable gear

(Note that, so far, there is no requirement for any avionics.)

Night VFR

☐ All of the above
☐ Approved position lights
☐ An approved red or white anti-collision light system
☐ If operating for hire, at least one landing light
☐ An adequate source of electrical energy for all installed electrical and radio equipment
☐ A spare set of fuses, or three spare fuses of each kind needed

(Note that there is still no requirement for any avionics.)

VFR

☐ All of the items required for day VFR, plus those required for night VFR if the flight is at night.
☐ Two-way radio communications system and navigational equipment appropriate to the ground facilities to be used.
☐ A generator or alternator of adequate capacity
☐ Sensitive altimeter, adjustable for barometric pressure
☐ Gyroscopic rate of turn indicator
☐ Gyroscopic bank and pitch indicator (artificial horizon)
☐ Gyroscopic direction indicator (directional gyro—DG—or equivalent)
☐ Slip/skid indicator
☐ A clock presenting hours, minutes and seconds with a sweep second hand or digital presentation

Now we need avionics, but the specific types are not spelled out. Just "appropriate" equipment is required.

MINIMUM EQUIPMENT REQUIRED IN CERTAIN AREAS

In addition to the requirements given above, other equipment is needed for flight in certain areas.

Airports With Control Towers

To operate at or in the vicinity of an airport with a US-operated control tower, you must have two-way radio communications equipment (FAR 91.87).

Terminal Control Areas (TCAs)

There are three groups of TCA. All flights, VFR or IFR, require a clearance from air traffic control (ATC) before entering, and ATC separation for flights within their boundaries. For flight within a TCA the following avionics requirements exist (FAR 91.90):

Avionics Required

Type of TCA	COMM Radio	VOR or TACAN	Transponder	Encoding Altimeter
Group I	Yes	Yes	Yes	Yes
Group II	Yes	Yes	Yes	No
Group III	Yes	No*	No*	No

*Provided two-way communication is maintained with ATC

Controlled Airspace

Flight within all controlled airspace above 12,500 feet within the contiguous United States, except that space within 2,500 feet of the ground, requires a transponder with an encoding altimeter, whether operating VFR or IFR (FAR 91.24).

All flight above 24,000 feet reqiring a VOR also requires distance measuring equipment (DME) (FAR 91.33).

Emergency Locator Transmitter (ELT)

In addition to the above requirements, FAR 91.52 requires most aircraft to be equipped with an ELT. The exceptions are jets, airliners, crop dusters, airplanes undergoing test flights and aircraft operating as trainers strictly within a 50-mile radius of their base.

Category II Operations

Certain aircraft may be operated in IFR to Category II minimums (roughly 100-foot ceiling and ¼ mile visibility or lower). Very specific requirements exist for this type of flight, not only in the equipment required, but also in the qualification of the crew, the creation and maintenance of a Category II flight manual and such. Details are spelled out in FAR 91.2 and FAR 91 Appendix A.

Transoceanic Operations

Single-engine aircraft operating across the Atlantic using Canada as a departure point must land first at Moncton, New Brunswick for an inspection by the Canadian Ministry of Transport. Avionics equipment required is:

☐ VHF radio communications equipment capable of using frequency 121.5 MHz (this frequency must be monitored throughout the flight)

☐ HF radio communication equipment with at least two of the appropriate frequencies (suggested: 5673 and 8888 kHz)

☐ Adequate radio navigation equipment, including two LF/MF receivers with a BFO (beat frequency oscillator) or CW (carrier wave) switch. One of these must have direction finding capability (e.g., an ADF)

Chapter 3
Avionics Discussion

Avionics found in general aviation aircraft can be divided into four broad categories: *communication, navigation, identification* and *control*.

COMMUNICATION

All modern communication between aircraft and the ground or other aircraft is made by voice. Standard communication frequencies are in the *very high frequency (VHF)* band, between 118.000 and 135.975, with frequencies moving up in 25 kHz steps—118.000, 118.025, 118.050, etc. up to 135.975. In current practice, only the 50 kHz steps are used—118.0, 118.05, 118.1, 118.15, etc. up to 135.95, except for certain high altitude IFR ATC channels. However, the .025 positions will come into increasing use in the future. To use all the 50 kHz frequencies, you need a 360 channel communication radio. To use the 25 kHz numbers, you need a 720 channel radio (Fig. 3-1). Older radios have 90 or fewer channels and can only use frequencies in 100 kHz steps (118.0, 118.1, 118.2, etc. to 126.9).

Military aircraft also use frequencies in the ultra high frequency (UHF) band, but these are beyond the scope of this book.

Long distance communications (primarily transoceanic or bush flying) use high frequency (HF) frequencies—those between 3 and 30 MHz (3000 and 30,000 kHz).

Emergency locator transmitters send out their signals on 121.5 MHz (the standard emergency frequency).

FCC Licensing Requirements

All radio transmitters installed in an aircraft must be licensed by the FCC. In addition to the transmitters used for communications, there are transmitters built into transponders, DMEs, weather radar and radar altimeters. Each transmitter must be FCC-licensed and specified on the aircraft's transmitting station license, which must be displayed in the aircraft. In addition, the operator of the equipment must have an FCC operator's license.

To apply for an FCC aircraft radio station license, you use FCC Form 404. To renew, you use FCC Form 405B. Applications are sent to FCC, Box 1030, Gettysburg, PA 17325. Operator's licenses require an application on Form 753 (755 if you are an alien), and are sent to Box 1050, as above.

When you buy an aircraft, you must obtain a new aircraft station license. The aircraft's existing license, if any, is valid for 30 days. If the aircraft is new, you may operate the equipment for 30 days before a license is issued. Since the FCC generally takes longer than 30 days to do *anything*, time is of the essence!

Ancillary Communications Equipment

Even the most professional of pilots can be bogged down by poor equipment, and even the best of radios can be spoiled by poor installation, malfunctioning antennas, faulty mikes, and such. Let's assume your radio is fine. The antenna is the right type. The wiring all checks out. How can you go better than that?

One Secret is Sidetone

There is a little electronic item called *sidetone*. Sidetone is what you have when you hear your own voice through the headphones as you transmit. Radio announcers often wear headphones, so they can hear exactly how they sound. Or they cup their hand behind an ear when they speak. Have you noticed how some people's voices change completely when they get on the telephone? Same reason. Telephones have sidetone, and when you can hear how you sound, you tend to modify your output to make it sound more mellifluous. You can bet that a pilot who's been listening to himself for a while on sidetone sounds pretty good.

Use a Headset

If you don't now wear a headset, borrow one and try it next time you fly. It's possible you may not be able to hear yourself,

Fig. 3-1. Narco COM 120 720-channel COMM displays every digit in the frequency (courtesy Narco).

because you may not *have* sidetone available, in which case the exercise is futile. If you have an audio switching panel, set the appropriate COM switch on "phones." The sidetone is a function of the transmitter on many sets—some of the signal is sent over the headphones, or it is processed through the audio control panel. But some older sets don't have it.

A Noise-Cancelling Mike Works Wonders

One of the most important things to have for good sound transmission is a noise-cancelling mike. This mutes loud background noice, while taking your voice signal at full strength. Some people don't know how to use a noise-cancelling mike. They hold it a few inches away from the lips. This is wrong. The correct technique is to hold it very close to—even touching—the lips, at the *corner* of the mouth. This is because the mike diaphragm receives background noise through holes on both the back *and* front sides of the mike, and your voice only through holes on the front side. The diaphragm detects those sounds which come through *both* front and rear sides simultaneously (i.e., background noise) and cancels them out, leaving just your voice to vibrate the diaphragm. Since your voice, using correct technique, only comes through the front holes, it is transmitted. If you hold the mike too far away, some of your voice will sneak in the back way and will get cancelled out by this feature. And holding the mike at the corner of the mouth reduces sibilance and P-pops—the loud pop sound you get when you say a P right into a microphone. Also, be sure you aren't covering the noise-cancelling holes with your hand.

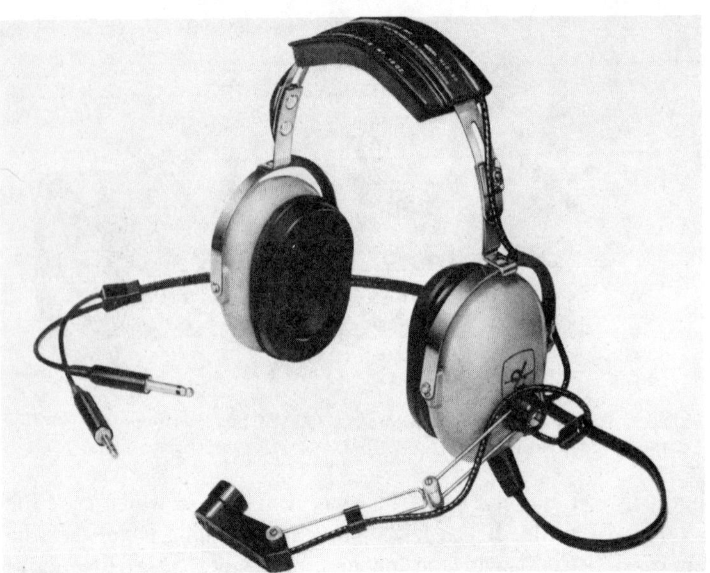

Fig. 3-2. David Clark H10-30 headset is ideal for use in high-noise environments (courtesy David Clark).

Put Your Hand-Held Mike on Stand-By

I don't understand why some pilots use hand-held microphones. Your hands are to fly the airplane, not to hold a mike! When you're flying you need one hand for the control wheel, throttle, trim control, and the other one for things like charts, switches, radio-frequency changes and so on. Why should a microphone be added to the list? If you have a hand-held mike, you've *got* to hold it in your hand—you can't crook it on your shoulder like a telephone when you're writing something. Some hand-held mikes are so badly designed you have to look at them first to figure which way you should hold them to put the words into the little holes! Hand-held mikes should be strictly for stand-by purposes. What you need is a boom mike that hangs off something mounted on your head, preferably headphones or eyeglasses. The type of mike that mounts on a goose neck attached to the cabin wall is a poor alternative, since this requires you to hold your face in a fixed position when talking—something you can't do if you're searching for traffic at four o'clock, unless you're a contortionist.

What Type of Mike?

When you look through the ads, catalogs and brochures you see there is a variety of mikes available for use in aircraft. The most common are the *carbon, dynamic* and *electret* types. The carbon mike is generally standard equipment in most planes, and is the worst type to have in an aircraft! Its only features are low cost and ruggedness. Its chief drawbacks are that it provides only fair frequency respones, intelligibility and noise cancelling capabilities. The dynamic and electret mikes are much better, offering excellent frequency response and intelligibility. They are also more costly. So if you have a carbon mike in your aircraft, relegate it to back-up use when you get your dynamic or electret boom mike!

What Type of Headphones?

You can get headphones with heavy sound insulation pads for use in high noise environments (Fig. 3-2), and you can get very

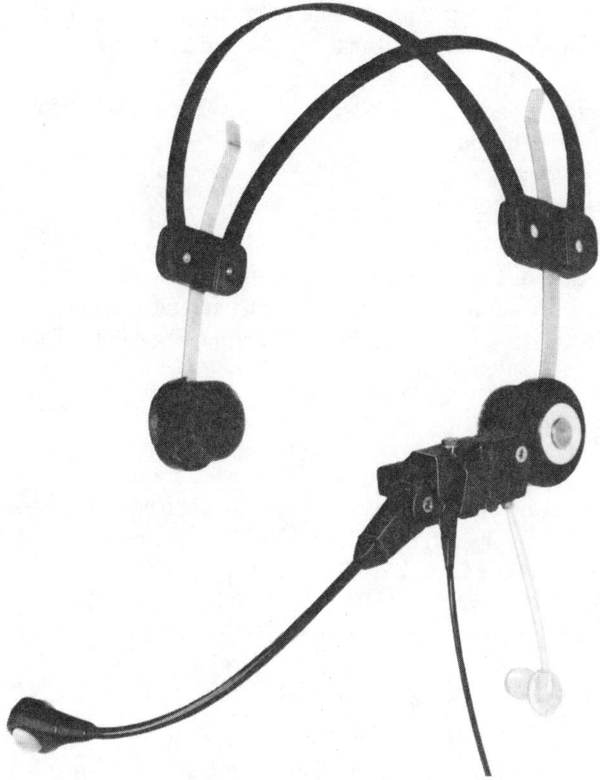

Fig. 3-3. Telex 5 × 5 Pro II headset is good for use in low-noise environments (courtesy Telex).

small lightweight phones that work well where the background noise isn't so bad (Fig. 3-3). So the first thing to decide is how much protection against noise you need.

Two types of headphone are available, *magnetic* and *dynamic*. The magnetic version provides a satisfactory signal quality for use in aircraft, but the dynamic, which is more expensive, produces a significantly better signal, and is recommended.

Headphones work in a manner very similar to their microphone counterparts, but in reverse, taking the electrical signal coming from the radio receiver and translating this into soundwaves by moving a diaphragm. Of course you need *aviation* headphones. Ordinary stereo phones won't do—many of them don't have the strength in their diaphragms, often made of paper, to put up with the heavy use they get in the cockpit.

One advance in hearing technology is the custom-moulded earpiece. This is a plastic plug moulded to your own inner-ear canal, attached to the headphone element. It thus transmits the sound to you both by normal air pressure and by conduction. Because of the personalized nature of the earpiece, it fits comfortably, and can be used to hold a very lightweight boom mike without a headband.

Airborne Telephones

Telephone coverage in the United States is not yet complete. There are still large areas of the midwest and west that are not covered. However, coverage in the east and far west is good, and an airborne telephone can be had for under $2,000. There are twelve channels available.

NAVIGATION

If the barnstormers of the 1930's were around today, they'd have to learn a whole new language just to get from point A to point B. Today's "alphabet soup" of navigation aids includes VOR, DME, TACAN, VORTAC, ILS, GS, CDI, OBS, ECDI, HSI, RMI, DVOR, RNAV, VNAV, ADF, and MB. Confused? Don't be. We'll take a look at each and discuss their functions and use.

VOR/DME

The primary civilian method of navigation in the United States is *VOR (VHF omnidirectional radio range)*, also known as *omni* or *omnirange*, combined with *distance measuring equipment (DME)*. The military uses a similar system called *TACAN (tactical air navigation)*. Most VOR and TACAN stations are combined, in which case they are called *VORTACs*. TACAN offers both an omnirange feature and a distance measuring capability. Civilian aircraft use the TACAN's DME transmitter and the civilian VOR, while military aircraft use the TACAN's omnirange and DME units. Some VORs have no TACAN, and thus no DME capability. And still others have no TACAN omnirange, but they have the TACAN DME component, in which case they are called VOR-DMEs. Got it?

VOR operates in the VHF band between 108.00 and 117.95 MHz. DME operates in the UHF band. It doesn't matter what the DME frequencies really are, since each one is paired with a VOR frequency. Thus when you tune a separate DME receiver, you tune it to the VOR frequency indicated on the chart and it secretly tunes in a UHF frequency for you. TACAN frequencies are referred to as channel numbers for military purposes, and there is a fixed relationship between each channel number and each VOR frequency.

Some military bases have only a TACAN transmitter. You can't use the omnirange component of this with your VOR receiver, but you *can* use the DME component. You simply tune the DME to the paired VOR frequency for that channel and you get distance from the TACAN on your DME. Most charts show what the VOR frequency "would be" for that channel number to facilitate this use.

Instrument Landing System (ILS)

Compatible with VOR/DME's use for primary cross-country navigation is the standard precision approach aid, the *Instrument Landing System (ILS)*. This receives its directional (*localizer— LOC*) signal in the same frequency band as VOR, but on special assigned ILS frequencies. A *glideslope (GS)* signal is received on a special UHF receiver, with frequencies paired to the VHF LOC frequency. The tuner that channels the LOC automatically channels the GS

Some ILS transmitters also send a DME signal, and are called ILS/DMEs. The DME tuner is set to the ILS frequency to receive this.

VOR/ILS Equipment

To use VOR and the ILS localizer you need a VHF receiver that receives on 108.00 to 117.95. Older units received only in 100 kHz steps, from 108.0 to 117.9, and are still quite acceptable. In

Fig. 3-4. Early Narco CDIs. VOA 3 (left) had VOR only, VOA 4 (middle) had VOR/LOC, and VOA 5 (right) had VOR/LOC/GS (courtesy Narco).

addition you need a VOR indicator and a converter—a device that converts the electronic signal being transmitted to a signal that can be used by the indicator. Most indicators have the converter built in.

VOR/ILS CDI

The VOR *course deviation indicator (CDI)* in its basic form consists of a simple left/right needle, a to/from indicator and an *omni-bearing selector (OBS)*, as shown in Fig. 3-4.

A more modern version is the *electronic course deviation indicator (ECDI)*, pioneered by Bendix, which has no moving parts. Needle movements are replaced by *light bars* illuminating, while OBS and to/from readings are presented digitally (Fig. 3-5).

The same CDI is used for ILS readouts, with the VOR needle or light bar being used for LOC readouts and a horizontal needle or vertical light bar to show GS deviations.

Horizontal Situation Indicator (HSI)

A much better system over the old CDI is that used in the *horizontal situation indicator (HSI)* (Fig. 3-6). This combines the VOR or ILS LOC information with magnetic heading data presented either with a directional gyro or a gyro-slaved magnetic compass. The HSI offers these benefits:

☐ Natural sensing of aircraft position relative to VOR or ILS courses at all times. You never have to turn away from the needle, as you do when flying outbound on an ILS or inbound on a back course localizer approach.

☐ No constant resetting of the omni bearing selector (OBS) when flying a VOR approach or a VOR hold to obtain correct needle sensing. Just set it once, and always turn toward the needle to get back on track.

☐ No need to figure reciprocals when flying VOR. Simply set the radial on the omni bearing selector (OBS) as you read it on the chart. The HSI takes care of the rest.

☐ Quickly shows if you drift off your heading during a precision approach, so you can correct *before* the needle goes off center.

☐ One less instrument to look at when flying IFR.

☐ No more confusing "to" or "from" flags. In a conventional VOR CDI, the flag can say TO when you are flying away from the station and FROM when you are flying toward the station. The HSI doesn't even have a "to" or "from" flag. Instead it has a *station pointer* (still called a "to-from pointer"), an arrow that points to the station. Thus you can always tell where the station is.

Fig. 3-5. Bendix ECDI uses light bars and has no moving parts. Pilot is flying inbound on the 305 radial, and needs to correct to the left (courtesy Bendix).

Fig. 3-6. Narco HSI 100S slaved horizontal situation indicator reveals all (courtesy Narco).

☐ Positive indication of whether or not you have reached a radial you are to "report crossing" or to intercept.

☐ Exact depiction of crab angle required to stay on track in a crosswind. Tells you how much crab you are holding prior to landing when you're doing an ILS approach.

☐ If you've never flown an HSI, you're in for a treat. Once you've flown it, you'll wonder why it's done any other way. All airliners and business jets use an HSI as part of their flight director systems. And now there are enough HSIs on the market that you can probably afford one if you can afford an airplane.

VOR Radio Magnetic Indicator (RMI)

Another good presentation of VOR information is with an *RMI*. It does not work on an ILS localizer, however. The RMI superimposes a "station pointer" needle on a directional gyro or slaved compass card. You read the aircraft's heading at the top at all times, and the VOR needle points to the station, giving you your relative and magnetic bearings to and from it at all times. You read your course to the station under the head of the VOR needle, and your radial from it under its tail. Most RMIs also have a needle for the *automatic direction finder (ADF)*, or a selector that enables you to choose whether the needle should point to a VOR or ADF station. More about ADF shortly.

Digital VOR

Many of the new NAVs on the market have a useful feature which was not available anywhere only a few years ago—digital VOR readout (Fig. 3-7). Digital VOR gives you a low-cost RMI (again—no ILS information). This means that you always have available to you your VOR radial or bearing TO the station. The readouts are presented to you digitally rather than as a needle reading, as on a conventional CDI. So if you are magnetically due west of a VOR, the DVOR would read 270% FROM and 090% TO the station. This can be a very useful feature.

I have a Davtron DVOR in my Comanche. This instrument is hooked up to both NAVs, a Narco Mark 12 driving a Narco HSI 100S and a Narvo NAV 122 self-contained NAV unit. I use it constantly, but I don't often use it to navigate to or from a station. However, it is useful when I have a routing "direct the VOR." In this case, I simply snap the reading to TO on whichever VOR I

Fig. 3-7. Davtron 902A digital VOR RMI is well worth the money (courtesy Davtron).

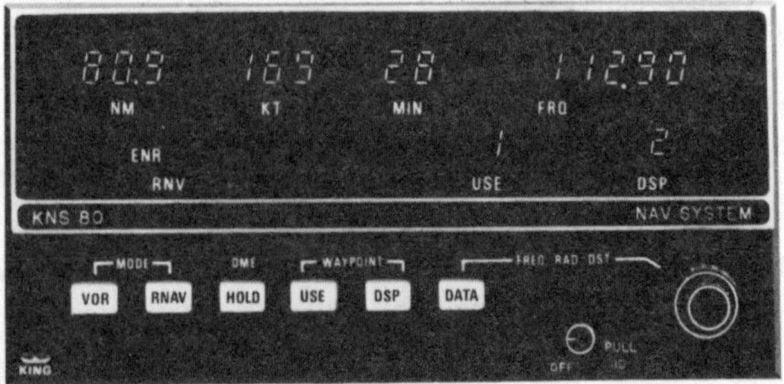

KNS 80 INTEGRATED NAVIGATION SYSTEM (VOR/DME/RNAV/ILS)

Fig. 3-8. King KNS 80 200-channel NAV, DME, GS, 4-waypoint RNAV and VOR converter is the dream of many pilots, including the author (courtesy King).

need, number one or number two, and turn to the heading shown. I just keep turning until my heading and the DVOR readings are the same number. Then, if I don't feel too lazy, I reach out and set the OBS on the VOR and fly the needle. Otherwise, I just fly the number on the DVOR as my heading. This automatically compensates for drift. However, I fly in on a curved line if there is wind.

DVOR Fluctuations

One thing about DVOR—expect a small amount of fluctuation in the readings. For example, it is not uncommon to see the a number that should be constantly decending—090°—080°—070° (such as when you are passing a station to your left)—occasionally the needle readings on your regular omni instrument. The numbers should be within 2% of each other.

Area Navigation (RNAV)

Area navigation (RNAV) is one of the most useful aids to flying cross-country available today (Fig. 3-8, 3-9). RNAV's primary function is to enable a pilot to electronically relocate, for his or her own purposes, and *VORTAC* (a combined VOR and TACAN station that gives both VOR and DME signals) to any location within a range of up 200 miles. The pilot can then navigate on this within normal line-of-sight range, which will depend on altitude.

Thus, with RNAV, by relocating VORTACs to suit your desired route at the touch of a button or the twist of a knob, you can fly in a straight line across the United States. No more zig-zagging to overfly VORs. You can navigate directly to your destination airport, or to any geographic location within range of a VORTAC as simply as you now navigate with VOR and DME.

With an FAA-approved installation, you can make straight-in IFR approaches at airports that don't have straight-in approach aids, such as ILS, LOC, LDA, SDF, NDB or VORTAC on the airport. The FAA has created many RNAV approaches, and is adding more all the time. You can also monitor published ADF approaches on your VOR, if there's a VORTAC nearby. You can navigate straight to the outer marker for an ILS approach—many approach charts show the RNAV coordinates of the outer marker. You can fly off airways, thus avoiding heavy-traffic areas. You can pick courses that take you safely around danger and restricted areas. You can, with an extra component called *vertical navigation (VNAV)*, set up glidepaths to give you an efficient descent line from cruising altitude to your destination, to the traffic pattern, or to *minimum descent altitude (MDA)* on a VOR or RNAV approach.

Not All RNAVs Use VOR/DME

Some RNAVs work with *very low frequency (VLF)* navigational stations, or an *inertial navigation system (INS)* that is contained totally within the aircraft. These are mostly found in military or

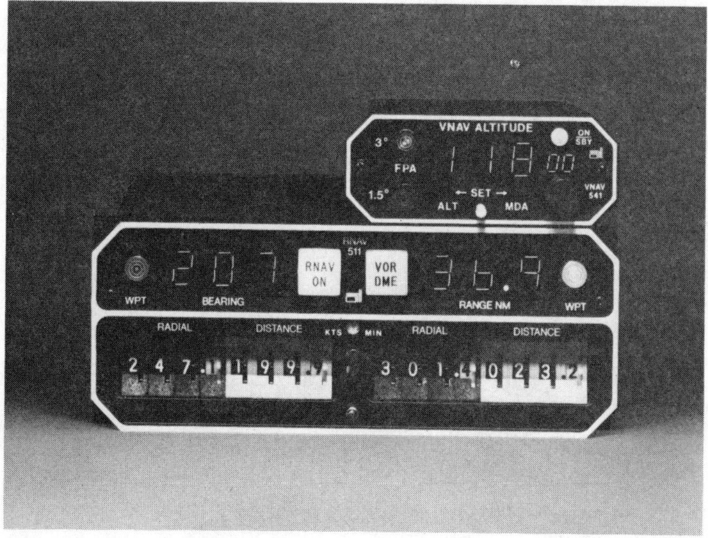

Fig. 3-9. Foster Airdata RNAV 511 and VNAV 541 work with virtually all NAVs and DMEs (courtesy Foster Airdata).

specialized aircraft, corporate jets and airliners, and are beyond the scope of this book.

Phantoms and Waypoints

With RNAV you can establish a *phantom* VORTAC station—called a *waypoint*—at any position within signal range of the VORTAC's actual location along any one of its radials. With most RNAVs, distances and bearings are selectable in increments of tenths of degrees and tenths of miles—e.g., 121.8° at 11.3 nautical miles (nm).

The RNAV displays distance to or from the selected waypoint either on readouts built in to the RNAV control panel or on the associated DME unit. Some RNAVs will give groundspeed and time-to-waypoint, too. Course deviation may be displayed on the RNAV's own readout and/or on the associated VOR/LOC course deviation indicator or HSI.

RNAV Isn't a Radio, It's a Computer

RNAV is not a radio *per se*. It is a computer that takes VOR and DME signals from your receivers and processes them to provide course and distance information for the pilot. The computer takes the airplane's *real* VOR bearing and DME distance (where you are *now*) and compares this with the location of the waypoint you have selected (where you want to go), also expressed as a VOR bearing and DME distance. By ingenious trigonometric calculations, performed by a microprocessor in modern units, the computer works out what the VOR needle and DME readout should say if the VORTAC really existed where you have put it.

Similar but Different

You use the VOR steering (course deviation) needle and to/from pointer almost the same as with regular VOR when flying RNAV. The waypoint becomes the "station." There is a slight difference in interpretation with most RNAVs, however. This is because the steering needle measures *distance* off the centerline of the course when you're using RNAV with most units. When you're flying normal VOR, the needle displacement measures your *angular* displacement. For example, a one dot deflection in pure VOR navigation on most VOR ilndicators means you are two degrees off course. At 60 miles from the VORTAC, this indicates you are two miles off the course centerline, and at 30 miles out you would be one mile off the centerline. A one dot deflection when you're flying RNAV means, on most indicators, you are a half-mile off the centerline, regardless of how far you are from the station. So, in RNAV mode, the VOR needle measures your *distance off track*, not your angle. Otherwise, you treat it as a regular VOR needle. Your to/from pointer operates just as it does with a VOR. When you go over your waypoint, you get a "station passage."

A small annunciation of the word "RNAV" appears on the face of some navigation instruments so that you know you're not using raw VOR/DME data when you're on RNAV.

Sensitive About Approaches? So is RNAV

When you're doing an approach with RNAV, you need a more sensitive needle, as with an ILS. Some RNAVs have an *approach* setting that makes the needle displacement twice as sensitive. Each dot off could mean a quarter of a mile displacement on such units, just as localizer needle deflection is more sensitive than that in VOR navigation.

Multi-Waypoint RNAVs

Some RNAVs enable you to preselect several waypoints, making for easier navigation, since you can preprogram your route beforehand, easing your workload during the flight. Such units can be easily reprogrammmed in flight if it becomes necessary to change the routing.

Compatibility of Units is Essential

When installing RNAV, you must have a compatible VOR NAV system and a compatible DME system. The accent here is on the word *compatible*, because not all RNAVs will work with all NAVs and DMEs. Quite the contrary—most RNAVs will only work with certain *specific* NAV/DME systems.

As a rule of thumb, most RNAVs of a given manufacturer will work with the more recent NAV and DME system of the same make, and not with competitive makes.

FAA Approval Required for IFR Use

Your RNAV must be approved by the FAA for use in IFR. Then you can simply file "RNAV direct" and carry out RNAV approaches. FAA Advisory Circular AC 90-45A *Approval of Area Navigation Systems for use in the US National Airspace*, Appendix

A, section 3 *Testing Procedure for Equipment Provided for use Under IFR* and Appendix B *Procedure for Obtaining FAA Data Approval by STC or FAA Form 337* give the details. It is available free from U.S. Department of Transportation, Publications Section, M 443.1, Washington, DC, 20590. Essentially the requirements, which apply to each individual airplane for which approval is sought, are as follows:

☐ The equipment must first be bench checked (this will probably have been done by the manufacturer, which will suffice);

☐ The electrical installation in the airplane must be approved;

☐ A functional ground test must be carried out;

☐ The equipment must be flown to demonstrate the accuracy of the system. This flight test must be carried out under VFR, solely by reference to the RNAV system and flight instruments, with a safety pilot on board. Ideally it should be carried out under ground radar surveillance, and should cover FAA-approved segments of an RNAV route, an RNAV terminal area procedure and an approach procedure.

Until such time as the tests have been completed to the FAA's satisfaction, each installation must be placarded "Area Navigation limited to VFR use only."

Vertical Navigation (VNAV)

Vertical navigation can be added to some RNAV installations. Some types give a digital presentation of intended altitude to make good, while others may be hooked up to a glideslope-like pointer on the HSI or flight director, but these are not usually found in light aircraft.

Automatic Direction Finders (ADF)

The *ADF* is a very useful navigational aid, particularly in under-populated areas. It operates in the low/medium frequency (LF/MF) bands, between 200 and 450 kHz. It also is of great help when doing an ILS approach, since most facilities include a compass locator beacon either at the outer or middle markers, providing you with directional guidance to those locations. ADFs may also be used on standard AM broadcasting stations. ADF tuners receive between 200 and 1750 kHz. With an ADF in your plane you get not only useful navigation guidance, but also the ability to listen to AM radio broadcasts in flight (Fig. 3-10).

Fig. 3-10. Narco ADF 141 offers crystal tuning and a unique "no signal" light on the ADF indicator if the station goes out (courtesy Narco).

Fig. 3-11. An ADF digitizer is the answer if you have an old analog-tuned ADF you want to upgrade for very little money (courtesy Davtron).

ADFs used to have analog tuning dials just like a household radio. More recently crystal controlled ADFs have become available, offering positive tuning. If you have the old type of analog tuner, you can obtain a digital indictor for a few hundred dollars that will indicate the exact frequency tuned in large illuminated digits (Fig. 3-11).

ADF navigation information is presented on a dial that looks like a compass card, with the top of the card representing the aircraft's nose and the needle pointing to the station (Fig. 3-12). If the needle points straight up, the station is straight ahead. If it points to 3 o'clock, the station is off your right wing. Older ADFs had fixed azimuth cards, making it necessary to do complex sums in your head to figure magnetic bearings. The newer ADFs feature a rotatable card. You set your heading under the lubber line at the top and read off your magnetic bearing under the needle's nose. Your "radial" is read under the needle's tail.

ADV/RMI

The RMI (see VOR/RME) can also be used to show ADF information. This takes the mystery out of navigation by ADF, and is strongly recommended. The azimuth card rotates automatically as the aircraft turns, and always shows the magnetic heading at the top.

Marker Beacon (MB) Receiver

The 75 MHz marker beacon is a vertical signal sent at specific locations to identify a positive fix. They are most commonly found

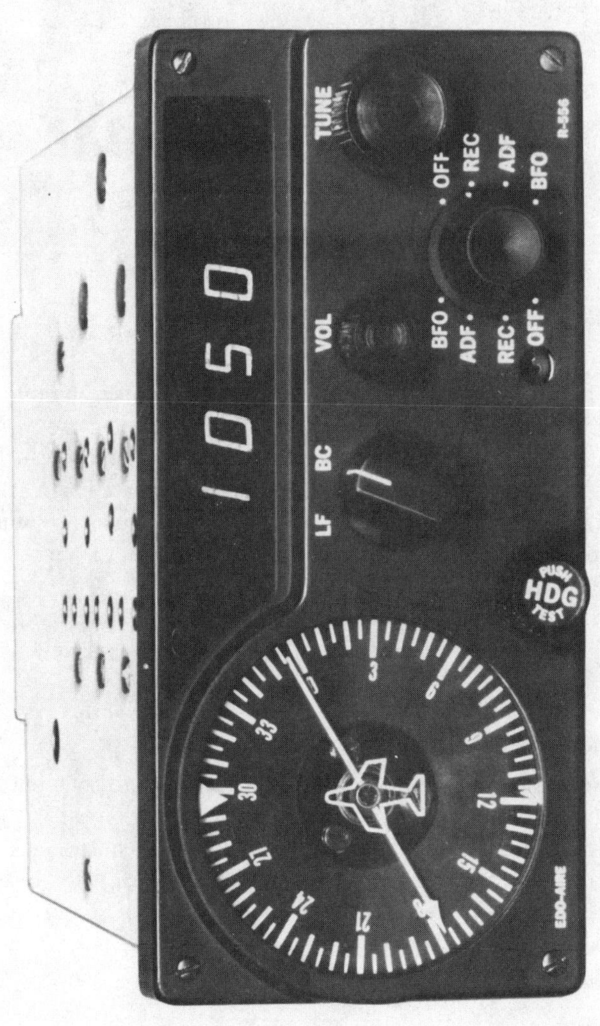

Fig. 3-12. Edo-Aire R 556 ADF looks like it's crystal tuned, but it's not. It has analog tuning and digital readout (courtesy Edo-Aire).

Fig. 3-13. This is the Narco MKR 101 marker beacon indicator (courtesy Narco).

in the ILS, where they are used to denote the outer, middle and inner markers. Some fan markers are also found to denote certain navigational fixes. The signal received is both aural and visual. The aural signal consists of a tone in morse code of a certain pitch, depending on the type of fix. The visual signal is usually a three light indicator on the panel, blue for outer markers, amber for middle markers and white for inner or fan markers.

Many marker beacon receivers and indicators are built into other instruments, most commonly audio control panels (Fig. 3-13).

Audio Control Panels

With the plethora of radio aids available to today's pilot, a switching panel is useful, to control which radio signal is to go where, between pilot and copilot headphones or cabin speakers. Many of these include an audio amplifier and a marker beacon receiver, as just mentioned. These units also select which transmitter is being used (Fig. 3-14). Some also incorporate a cabin intercom.

Weather Avoidance Systems

There are two electronic ways of avoiding thunderstorm activity in an airplane in flight—weather radar and the Ryan Stormscope. In fact the Stormscope even works on the ground. Radar has traditionally been only available in twins, but in the last

Fig. 3-14. Collins AMR 350 audio control panel includes amplifier and marker beacon lights. (courtesy Collins).

few years has become usable in single-engine aircraft as well. There are three ways of mounting the antenna in a single. One is in a pod on the leading edge of the wing or suspended below it (known as "the bomb"). Another is to use a very shallow antenna and put it right inside the wing leading edge. And the third is to mount it in the nose beneath the prop.

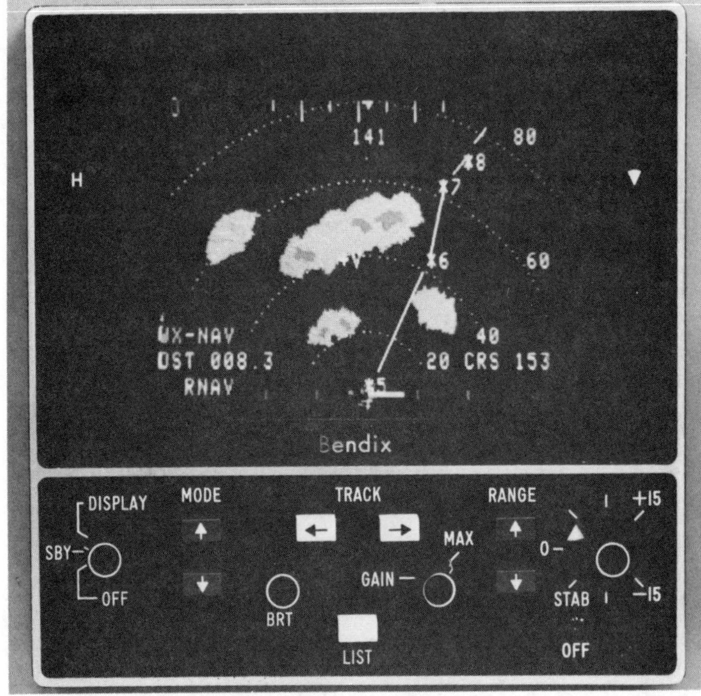

Fig. 3-15. The latest Bendix Colorvision digital radar will not only display weather, you can overlay your route on it too, if you have a Bendix 200 system (courtesy Bendix).

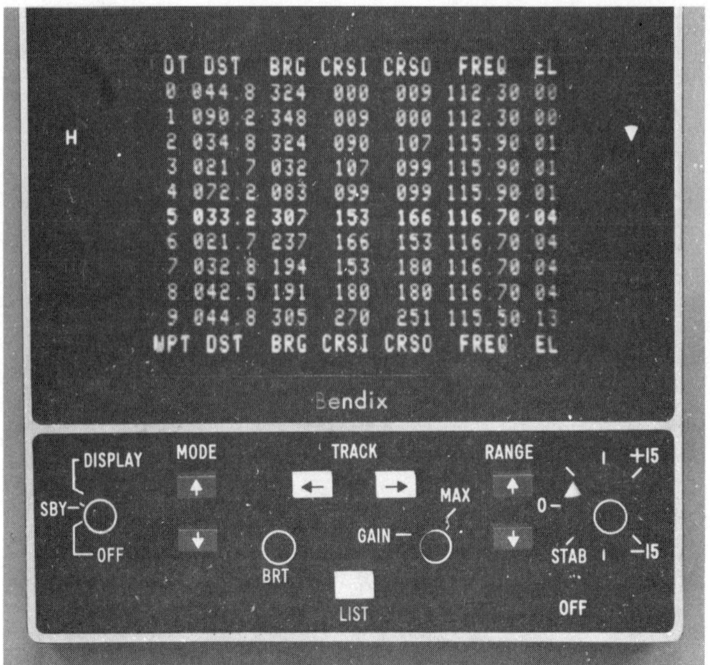

Fig. 3-16. You can display your RNAV waypoints (courtesy Bendix).

Modern radar is available in color, giving a much better indication of nasty weather. These days you can also use the screen to display checklists, RNAV waypoints, even a map of where you're going with the weather depiction overlaid (Figs. 3-15, 3-16, and 3-17).

The Ryan Stormscope is a system that works a bit like an ADF. Weather radar detects an echo caused by precipitation. The Stormscope detects electrical energy —lightning flashes. Both use a cathode ray tube (CRT) for display. Weather radar scans the whole picture many times a minute, and thus gives a constant indication of what's happening. Stormscope displays and stores data on a CRT in the form of dots on an azimuth screen (even behind you—weather radar won't do that). When 128 dots have been displayed, the oldest one is deleted and the new ones go on. Thus, if there's not much activity, the screen doesn't change much. If you should happen to turn in the interim, the screen won't reflect the relationship between your new heading and the old dots. Stormscope is now FAA approved for use as a weather avoidance system where one is required.

Radar Altimeter

The *radar altimeter* (Fig. 3-18) is one of the devices needed for Category II operations. It measures your height above the ground by a radar pulse. It starts working below about 2,000 feet. On some units you can set your decision height so that it will tell you when you get there. The unit is also valuable as a ground proximity warning indicator.

INDENTIFICATION — TRANSPONDER AND ENCODING ALTIMETER

A *transponder* (Fig. 3-19) is now required for all operations within controlled airspace above 12,500 feet, VFR or IFR, except when within 2,500 feet of the ground (FAR 91.24), and in certain terminal control areas. When a transponder is required, it must include an *altitude encoder* that will respond to "mode C" interrogations by sending the plane's altitude in 100 foot increments. This information will then display on the controller's radar screen, along with computed speed and aircraft identification, derived from the transponder code.

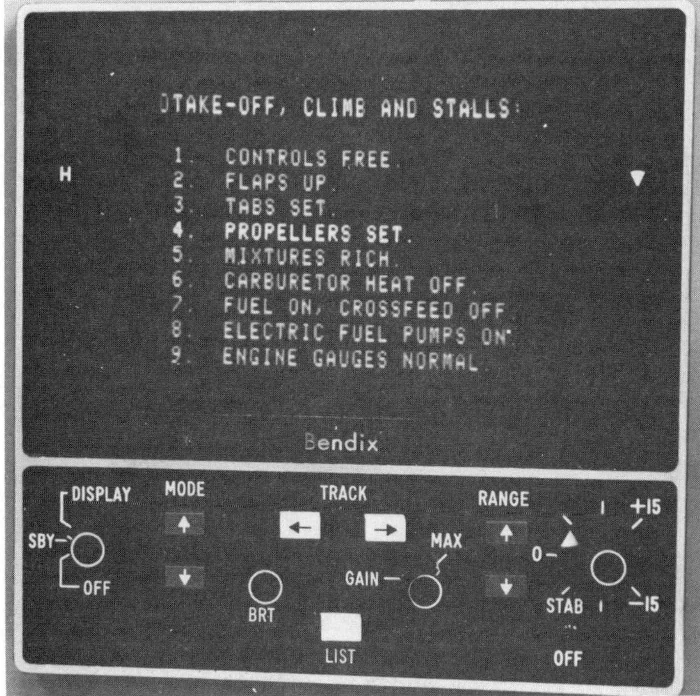

Fig. 3-17. You can display checklists (courtesy Bendix).

Fig. 3-18. King KNI 416 radar altimeter. Note decision height marker, here showing at 200 feet. The small light marked "DH" illuminates when you get there (courtesy King).

Not all transponders have an encoder attached. Even though the transponder has an "ALT" or "Mode C" position, this will nto work if there is no encoder. Some encoders are a part of the airplane's altimeter, others are "blind encoders," i.e., a unit somewhere in the bowels of the airplane with no visual readout. It is possible to buy an indicator that will display the altitude that the encoder is sending out, for verification within the cockpit.

All transponder *must* be TSO'd. If your aircraft has one that isn't, it is illegal to use it. Some non-TSO'd transponders can be modified, while others can't. If yours can't, remove it and sell it abroad, where non-TSO'd transponders are still legal.

Because of the need for a transponder, many aircraft carry two.

CONTROL—AUTOPILOTS AND FLIGHT DIRECTORS

Most *autopilots* (Fig. 3-20) are sold along with the aircraft as an optional extra. Autopilots can operate about a single-axis (ailerons only), two-axes (ailerons and elevators) or three-axes (ailerons, elevators and rudder). The simplest form is the

Fig. 3-19. Edo-Aire RT 887 "Squareponder" fits in a standard 3-inch instrument hole, a useful feature on a crowded panel (courtesy Edo-Aire).

Fig. 3-20. Brittain Nav-Flite II Autopilot includes heading hold and NAV couplers (courtesy Brittain).

Fig. 3-21. Some autopilots require their own DG (courtesy Brittain).

single-axis, often called a wing-leveler. Most autopilots also offer navigational couplers. A coupler takes VOR or ILS signals and makes the airplane fly along the course selected by the pilot.

Another useful feature on some autopilots is an altitude hold. This can be simply a button that you push when you want to hold the altitude where you are. Exotic models enable you to dial in a desired altitude, and the airplane climbs there, levels off, and asks you how you want your coffee.

Some autopilots work off the artificial horizon, turn and bank/turn coordinator and/or directional gyro (Figs 3-21, 3-22). This could mean a need to change these instruments if you are adding the autopilot later. If this is the case, it might be worthwhile keeping the old instruments as standbys. Redundancy is all important in airplanes, so if you are getting an electric DG with your autopilot, keep your old air-driven model as a stand-by, or vice versa.

Most, but not all, HSI's can accept autopilot inputs. If you have an HSI and are considering an autopilot, or vice versa, make sure that they'll talk to each other.

A *flight director* (Fig. 3-23) processes navigation and attitude information through a computer. Then it gives the pilot commands to fly the aircraft so as to achieve the desired flight path. It makes the task of flying the airplane on instruments very easy. The information is displayed through a director or horizon, either by "V-bars" or a cross-pointer. The pilot flies the airplane to satisfy the commands given by the director horizon and, thanks to the computer, the airplane will do whatever it has been programmed to do. For a go-around from an ILS approach, for example, the pilot, on rejecting the approach, pushes a button on the control wheel or throttle and the director commands an immediate pitch-up to climb out at the correct angle.

An important part of the flight director system is an annunciator that tells the pilot how the director has been programmed. There have been several accidents caused by the pilot flying commands of an incorrectly programmed flight director, and driving right into the ground.

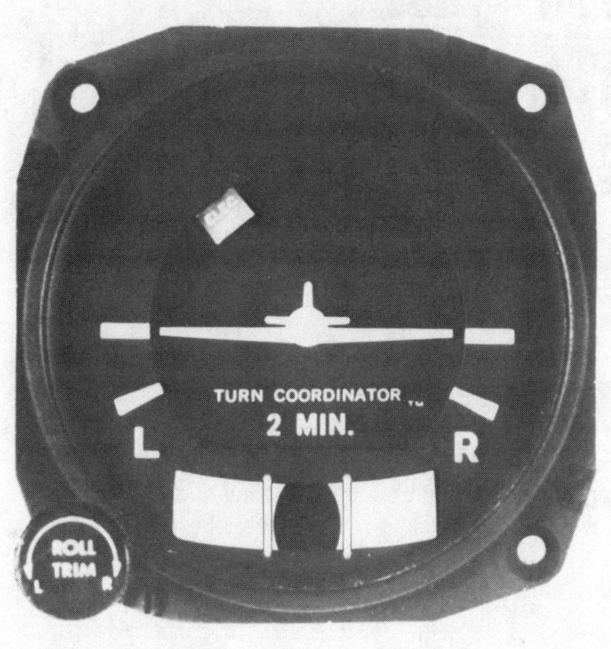

Fig. 3-22. Some require turn coordinator (courtesy Brittain).

Fig. 3-23. This is the Edo-Aire Mitchell Century 4 autopilot and flight director system; (courtesy Edo-Aire).

ANTENNAS

One of the worst problems I've ever encountered with avionics occurred in recent months with my Comanche. All of a sudden the VOR needles started waving at me. I know we were good friends, but I would have preferred proper decorum at all times—ie., a nice firm needle indicating what I wanted to see. Much trouble-shooting finally zeroed the problem in to the antenna installation. I had put in a new VOR "towel-bar" antenna, supposedly the best type you can get. However, when it was installed, the FBO had failed to make the connection between the antenna and the fin a positive electrical one. Paint that should have been scraped off was not, and that was part of the problem. It also turned out, when we were replacing the antenna cable (on the third attempt at isolating the problem), that the old VOR antenna lead running inside the leading edge of the fin had been partially drilled through when the original ADF antenna mount had been installed on the vertical fin. This must have been almost 15 years ago. It had taken that long for the slight nick in the antenna lead to corrode enough to cause a short and a loss of signal.

So a good antenna system and first-rate installation are essential. The best radios will work very poorly if their antennas are poor. Proper selection, placement, shielding and mounting are all part of the process.

STATIC DISCHARGE WICKS

If you fly through rain very often, you should have *static discharge wicks*. These are small, taperlike devices mounted on the trailing edge of control surfaces. They serve to eliminate precipitation static by discharging the electrical buildup that the aircraft acquires when flying in these phenomena. Without them, your radios will sound terrible in weather—and could become unusable.

Chapter 4
What Are Your Avionics Needs?

The type of flying you do dictates the type of avionics package you should have. Avionics systems can cost as much as, or more than, the aircraft itself. So unless money is no object, careful selection is important. Let's look at what I suggest as the minimum radio packages you should have for various types of flying.

BASIC VFR

This is the puddle-jumping variety, in an old airplane, such as a Piper J-3 or Aeronca, an Ercoupe or maybe a Cessna 140. The very minimum requirement is for an ELT. The next thing you should have is a basic communications radio. It doesn't have to have a lot of channels. Maybe one of the old 90 channel jobs will do—or even an old Narco Omnigator. If you can get a 360 channel COMM, so much the better. Next comes some form of simple NAV. Some combo units put a COMM and a NAV in one box—the old King KX 150 was an example, or the Narco Escort 100. In some the receiver is shared between the COMM and the NAV, so you can't communicate and navigate at the same time. This is probably fine for this type of airplane.

If you've still got room, money and electrical power, I would definitely add a transponder at this point. I upgraded my transponder recently from an old Narco AT 6A to a new Collins TDR-950, and I sold the old unit to a friend for his Ercoupe for $100. He's tickled pink with it. It works fine, and he gets a lot better service from control towers than he used to.

Fig. 4-1. Even a lowly Cessna 152 can be fully equipped for serious VFR or even mild IFR. Avionics by ARC (courtesy Cessna).

MILD VFR CROSS-COUNTRY

Now you're probably into a Tri-Pacer, a Stinson or an old Cessna 172. You don't mind making the odd 100 to 300 mile trip, when the weather is CAVU. You don't fly into any TCAs, and go to few airports with control towers. For this type fo flying I would aim for at least one NAVCOMM that will give you basic VOR and 360 channels of communication power simultaneously. For example, the old reliable Narco Mark 12 or the newer King KX 170 or 175 would be fine. A transponder is getting to be a must for this type of flying—they can be bought used for around $200.

SERIOUS VFR—LEARNING IFR

Here you are probably flying a newer airplane—maybe a recent Skyhawk or Cherokee. Two full 360 channel NAV COMMs are recommended. A transponder for sure. DME? Why not? You'll need a glideslope receiver if you're getting your instrument rating. And you should have an ADF and marker beacon receiver. If you hit many TCAs, get an encoder for the transponder. If money is no object, add an RNAV. The more you use it, the more you'll like it (Fig. 4-1).

MILD IFR

The minimum equipment for this type of flying, in my opinion, is all the stuff mentioned above, and make sure it *works*. An autopilot would also be a good idea. A digital VOR is strongly recommended. And I would definitely put in an HSI. It makes life so much easier.

Table 4-1. Suggested avionics for various types of flying.

TYPE OF FLYING YOU DO					
Avionics Item	Basic VFR	Mild VFR X C	Serious VFR Student IFR	Mild IFR	Serious IFR
ELT	Min	Min	Min	Min	Min
COMM 1					
90 ch	Yes				
360 ch	Nice	Min	Min	Min	
720 ch			Nice	Yes	Min
NAV 1					
VOR/LOC	Nice	Min	Min	Min	Min
GS 1			Min	Min	Min
XPDR 1	Nice	Yes	Min	Min	Min
ENCDR 1		Nice	Yes	Min	Min
DME		Nice	Yes	Yes	Min
ADF		Nice	Yes	Min	Min
MKR			Min	Min	Min
AUDIO		Nice	Yes	Min	Min
A/P		Nice	Nice	Yes	Min
HSI			Nice	Yes	Min
DVOR			Nice	Yes	Min
COMM 2					
360 ch		Nice	Yes	Min	Min
720 ch				NIce	Nice
NAV 2					
VOR/LOC		Nice	Yes	Min	Min
GS 2				Nice	Yes
XPDR 2				Nice	Yes
ENCDR 2				Nice	Yes
RNAV			Nice	Nice	Yes
RAD ALT			Nice	Nice	Yes
RADAR/ STORMSCOPE			Nice	Nice	Min
RMI			Nice	Nice	Nice
FLT/DIR				Nice	Nice
PHONE			Nice	Nice	Nice

SERIOUS IFR

Add to the mild IFR equipment a second transponder, dual glideslopes, and an RMI. Next: weather radar or a Ryan Stormscope (or both). If you can afford it, throw in a radar alimeter. And a flight director would be nice, too (Fig. 4-2).

Table 4-1 will help you interpret the need for various avionics for the type of flying being done. To use the table, find the type of flying you do *mostly* in the heading on the top e.g., basic VFR, mild IFR, student IFR, or whatever. The radios you should have, in my opinion, are listed down the left hand side. The table tells you whether I recommend that you *must* have an item (min), *should*

Fig 4-2. A Cessna 310 equipped for "serious IRF." Note radar, RNAV, RMI, flight director and all the other goodies. An ARC package (courtesy Cessna).

Table 4-2. Checklist of proposed avionics.

Avionics Item	Need	Desire	Price Range	Make	Model	Remarks
ELT	( )	( )				
COMM 1						
90 ch	( )	( )				
360 ch	( )	( )				
720 ch	( )	( )				
NAV 1						
VOR/LOC	( )	( )				
GS 1	( )	( )				
XPDR 1	( )	( )				
ENCDR 1	( )	( )				
DME	( )	( )				
ADF	( )	( )				
MKR	( )	( )				
AUDIO	( )	( )				
A/P	( )	( )				
HSI	( )	( )				
DVOR	( )	( )				
COMM 2						
360 ch	( )	( )				
720 ch	( )	( )				
NAV 2						
VOR/LOC	( )	( )				
GS 2	( )	( )				
XPDR 2	( )	( )				
ENCDR 2	( )	( )				
RNAV	( )	( )				
RAD ALT	( )	( )				
RADAR/	( )	( )				
STORMSCOPE	( )	( )				
RMI	( )	( )				
FLT/DIR	( )	( )				
PHONE	( )	( )				

Fig. 4-3. This Cessna Skylane could use some help (courtesy John Ferrara).

have an item (yes) or if it would be a good idea to have one, although it's not absolutely necessary (nice).

If you are equipping an airplane from scratch, this table will help you decide what to install. If you already have a plane, it will tell you what I think you ought to have in it (Fig. 4-3). If you are buying a used one it will tell you what I think you should look for.

If you are starting from scratch, use Table 4-2 to list your ideal minimum package.

Chapter 5
What's Wrong
With Your
Current Avionics?

If you·already have a plane, or you are buying a used one, you need to evaluate the existing avionics package in light of your needs, as established in the previous chapter (Figs. 5-1, 5-2). For this, we'll use another worksheet (Table 5-1) for you to list your current

Fig. 5-1. An upgrader's dream. Look at the old King KX 150B and the obsolete Narco Mark 10 in this Bonanza (courtesy John Ferrara).

Table 5-1. Avionics evaluation worksheet.

Avionics Item	Make	Model	Year Made	Does it work?	Should it be replaced?
ELT					
COMM 1					
90 ch					
360 ch					
720 ch					
NAV 1					
VOR/LOC					
GS 1					
XPDR 1					
ENCDR 1					
DME					
ADF					
MKR					
AUDIO					
A/P					
HSI					
DVOR					
COMM 2					
90 ch					
360 ch					
720 ch					
NAV 2					
VOR/LOC					
GS 2					
XPDR 2					
ENCDR 2					
RNAV					
RAD ALT					
RADAR/ STORMSCOPE					
RMI					
FLT/DIR					
PHONE					

Fig. 5-2. Here's Mooney crying out for aid. Note the glideslope antenna above the instrument panel (courtesy John Ferrara).

Avionics Item	Make	Model	Year Made	Does it work?	Should it be replaced?
ELT	Narco	ELT 10	?	Yes	No
COMM 1					
90 ch					
360 ch	Narco	Mk 12	1963	Yes	Eventually
720 ch					
NAV 1					
VOR/LOC	Narco	VOA 5	1963	Yes	Replace with HSI
GS 1	Narco	UGR-2	?	Yes	No
XPDR 1	Narco	AT 6A	1967	Sometimes	Yes
ENCDR 1	Aerosonic		?	Yes	No
DME	Narco	UDI 3	?	Sometimes	Yes
ADF	Motorola	T-12B	1963	Yes	Eventually
MKR	Narco	MBT 3	1963	Yes	Replace with audio panel
AUDIO	NO				
A/P	Piper	Auto II	1963	Sometimes	Yes
HSI	No				Add one
DVOR	No				Add one
COMM 2					
90 ch	Narco	Mk 12	1963	Sort of	Upgrade to 360 ch
360 ch					
720 ch					
NAV 2					
VOR/LOC	Narco	VOA 4	1963	Yes	Eventually
GS 2	No				
XPDR 2	No				
ENCDR 2	No				
RNAV	No				
RAD ALT	No				
RADAR/	No				
STORMSCOPE	No				
RMI	No				
FLT/DIR	No				
PHONE	No				

Table 5-2. Avionics evaluation worksheet (author's sample).

Table 5-3. Avionics replacement worksheet.

Item	Replace With
1. Immediately	
2. Eventually	

Table 5-4. Avionics replacement worksheet (author's sample).

Item	Replace	With	
1. Immediately			
NAV 1 VOR/LOC	Narco	VOA 5	HSI
XPDR 1	Narco	AT 6A	XPDR
DME	Narco	UDI 3	DME
MKR	Narco	MBT 3	Audio Panel
COMM 2 (90 ch)	Narco	Mk 12	Upgrade to 360 ch
2. Eventually			
COMM 1 (360 ch)	Narco	Mk 12	720 COMM
ADF	Motorola	T-12B	Digital ADF
A/P	Piper	Auto II	A/P
NAV 2			
VOR/LOC	Narco	VOA 4	NAV

avionics package, along with its positive or negative aspects and whether you intend to replace any. I have filled one out for you showing the situation I found when I bought my 1963 Comanche in 1976 (Table 5-2).

After you have filled out your sheet, make a separate list of the avionics that need to be replaced, along with any notes (Table 5-3). Table 5-4 shows how I filled out my "shopping list."

Chapter 6
Selecting
Your Radios

Note that when we have called for replacements, we have only named them generically. We haven't gone through the process of selecting a specific model yet. Now you have some idea of what you need, you have to find out what you're going to get. The primary consideration here, of course, is money. If you can afford it, buy new equipment from a factory-authorized dealer and have them install it all at once.

If you read the ads, and talk to radio shops and other pilots, you may have a reasonable idea of what you want in the way of brands and models. In this case, ask a couple of dealers to bid on installing these specific radios for you.

SELECTING NEW EQUIPMENT

If you don't really know what you should get in the way of brands, ask two or three dealers to make recommendations and bid on the job. Let's look at a couple of hypothetical situations. We'll assume that you have bought a new Cessna Skyhawk with no radios in it at a bargain price. The aircraft already has an ELT. You have evaluated your needs and have concluded that you would like a NAV/COMM 360 channel, a Transponder and encoder, and a DME, but you don't know what brands or models to get. This letter should help:

Dear________:

I have a 1981 Cessna Skyhawk. Its only radio is an ELT. I would like to install the following new Radios:

1 NAV/COMM 360 channel
1 Transponder and encoder
1 DME

Would you please recommend suitable models to me of good quality and low cost and give me your bid for supplying and installing them?

Yours truly,

If you send that to a couple of reputable dealers, you should get back a recommendation and a price for the total package installed. The price will be good for a specific period, such as 30 days.

After you have evaluated the bids, make a decision based on the radios being offered, the price, and the reputation and conveniences to you of the dealer. If the dealer is out of town, bear in mind you're going to have to leave the airplane with him for a few days, so you'll need transportation back and forth while he's got it.

Let's look at another example. Say you own a 1971 Piper Cherokee 235, and you want to upgrade some of the radios. Here's how this letter might look:

Dear________:

I own a 1971 Piper Cherokee 235. I want to upgrade my avionics package. The current equipment is as follows:

2 Narco Mark 12Bs
2 Narco VOA 8
1 Narco ADF 31A
1 Narco AT 6A (no encoder)

I want to replace the entire radio package with modern digital equipment, as follows:

2 NAV/COMMs
1 ADF
1 DME
1 Transponder and encoder

Would you please recommend suitable models to me of good quality and low cost and give me your bid for supplying and installing them, and indicate the trade-in allowance you would give?

Yours very truly,

USED AVIONICS

Many dealers will take your old radios in trade, and some companies have made a business out of the used-radio trade-in market. There is an active market in used avionics. The best source is *Trade-A-Plane*, which runs many columns of want-ads and a good selection of display ads in each issue on used avionics. You can subscribe to *Trade-A-Plane* by writing to them at Crossville, TN 38555. Most FBOs get it if you just want to see a copy.

There are two things to watch out for in buying used radios. One is the reputation of the seller. Make sure you get a warranty that's supportable. Buying a used radio "as is" means *you* will be responsible for getting it fixed when it fails after you've had it installed. The other is the cost of getting it fixed in the future. Some components of older radios are very expensive to replace, and it may just not be worthwhile buying one, whatever the price, because the repair bills will end up costing you more than a brand new radio.

Here are some of the more expensive problems that radios can have:

Type of Radio	Common Problems
Old COMMs & NAVs:	Frequency selector wafers; HF power transistors
Newer COMMs & NAVs:	Frequency synthesizers
NAV indicators:	Meters and flags
DMEs:	UHF tube, cavity assembly
Transponder:	UHF tube, cavity assembly
Old ADFs:	Loop antennas

UNDERSTANDING TAGS

Often when you buy a used radio (or any piece of equipment for an airplane) it will be tagged with a colored label, or simply be referred to as follows:

Yellow tag; Part has been overhauled and is approved for return to service. Tag will be dated and signed by the inspector who approved it. Watch out for old dates. Time has a tendency to spoil even fully overhauled equipment that has been sitting on a shelf.

"OHC:" "Overhauled and certified." The equivalent of a yellow tag.

Green tag: Part was removed in airworthy condition from aircraft. Has not been overhauled. Should also be signed and dated.

"Working when removed:" The equivalent of a green tag.

Red tag: Part has been rejected and is considered not airworthy.

"As is:" Caveat emptor.

If you buy a piece of tagged equipment, keep the tags in the airplane's log book for future reference.

MAKE SURE YOU GET EVERYTHING YOU NEED

A radio is not just a radio. It's also connectors, racks, leads, antenna couplers, converters, and the like. When you buy a new radio, all these things are neatly packaged in the box. What about when you buy a used radio? You may be lucky enough to get everything, and then again, you may not. Before you agree to purchase a used radio from far away, make absolutely sure that you're getting everything you need. The best way to find out what you'll need is to talk to the radio shop that will do the installation and find out what you are going to need. For example, when I bought my HSI 100, it was replacing a DGO 10. It was running off the same radio as before, a Narco Mark 12. I needed two things, a VOR converter and a new connector. The converter was necessary because the DGO 10 had its own converter while the HSI 100 didn't. The new connector was necessary because the DGO 10 and HSI 100 plugs were incompatible. So when I ordered the HSI to be shipped, I was able to make sure exactly what I was getting.

GETTING USED RADIOS INSTALLED

Part of the problem with buying a used radio is getting it installed. If you buy it through the mail, you're going to have to find a radio shop that will install it for you. Some dealers advertise that they will cheerfully install radios you bring in: others don't. Their attitude is, if you want a radio, they'll be happy to sell one to you, and install it. Their problem is if they install a radio that they don't know (i.e., one you bring in out of a plain brown wrapper), what happens after it's in and it doesn't work? You could say, "Well, it's your installation—fix it!" And the dealer could say, "Listen, I don't know where this radio came from. I put it in your airplane the way it's supposed to go, and it doesn't work. It's *your* problem. If you want *me* to fix it, my labor rate is $35 an hour!"

The solution is to develop an ongoing relationship with a radio shop whose work you trust and work out any arrangements about used equipment well in advance. The radio shop I use is Linden

Avionics at Morristown, NJ. The folks there have worked on my Comanche many times, and they know what's going on in it. I have had radios put in that they sold me, and radios that I brought in myself. They have removed radios for me to resell and trouble-shot my little problems. I feel good about using them, because I know that they are concerned about my plane and will do the right kind of job for me. Most recently they completely rebuilt my instrument panel, with very satisfying results.

As for used radios, I have bought and sold with good results from Connecticut Avionics and Aircraft, Inc. at Box 555, East Granby, CT 06026. They sold me the HSI to replace my aging Narco DGO 10. I traded my old ADF and DGO 10 for a never-used Narco HSI 100, with a reasonable cash difference. When we got the HSI to the shop, we found that the OBS didn't rotate all the way. I called CAA, since it had a 90-day warranty. They told me to send it to their instrument shop for fixing. I didn't want to wait that long, and I asked if they had anything else—such as a slaved HSI (the one I had bought was unslaved). It turned out they had a like-new HSI 100S. So for a modest increase in price I got what I really wanted, and returned the defective unit to their shop.

I had an old Narco UDI 3 DME that was driving me to the poor house. I decided I'd spent my last $300 on it, and obtained a new DME 190. I sold the UDI 3 to a friend for $100, on the strict understanding that it was "as is" and not working well when I removed it. He had it bench-checked and tuned for $125, and installed it in his Apache. It has since failed once, at a fixing cost of $170, and he says "I've got a working DME in my airplane and it cost me less than $300. But I did this once before with a UDI 4 DME in another airplene, and it cost me over $800 in one year." Buying as-is equipment means you get what you pay for.

Chapter 7
Additional Benefits
Of Upgrading

Apart from the obvious benefits of getting better radios that work more reliably and that do more for you, another key benefit of going to modern equipment is that it is often much lighter and uses much less power than the old equipment. Let's take a look at what happened as I gradually upgraded my avionics package over the years. We'll start with the system as it was when I bought it:

Original package, as installed at time of purchase

Function	Type	Load (amps)	Weight (lbs)
AUDIO	None	0.0	0.0
ADF	Motorola ADF T-12B	1.3	10.9
NAVCOMM 1	Narco Mk 12 360 ch VOA 5	11.0	13.0
VOR/ILS	Narco VOA 5	0.4	3.3
NAVCOMM 2	Narco Mk 12 90 ch VOA 4	11.0	13.0
VOR/ILS	Narco VOA 4	0.4	3.3
G/S	Narco UGR 2	0.2	3.1
MKR	Narco MBT 3	0.1	1.2
XPDR	Narco AT 6A	1.8	7.0
ENCODER	Aerosonic 3P	0.0	1.0
DME	Narco UDI 3	6.5	8.5
ELT	Narco ELT 10	0.0	3.7
Totals		32.7	68.0

You can see that the total weight of the package was 68 pounds and it took 32.7 amps from my 35 amp generator—not a very healthy situation. I started searching for an improved package. My first attempt produced the following combination:

Function	Type	Load (amps)	Weight (lbs)
Step One			
Delete:			
VOR/ILS	Narco VOA 5	0.4	3.3
MKR	Narco MBT 3	0.1	1.2
Add:			
Audio	King KMA 20	1.1	2.3
HSI	Narco DGO 10	1.5	4.7
ADF IND	Davtron 701	.6	0.4
IVSI/GS DIR	Teledyne SLZ 9541	0.1	0.0
New totals		35.5	70.0

This didn't help my weight any, and increased my load (Fig. 7-1). It was back to the drawing board.

Function	Type	Load (amps)	Weight (lbs)
Step Two			
Delete:			
XPDR	Narco AT 6A	1.8	7.0
Add:			
NAVCOMM	Convert 90 to 360 ch	0.0	0.1
VOR IND	Davtron DVOR 902A	0.0	0.4
XPDR	Collins TDR 950L	1.3	2.0
New totals		35.0	66.4

The weight was coming down, but not the electrical load. I gave it another try:

Function	Type	Load (amps)	Weight (lbs)
Step Three			
Delete:			
CME	Narco UDI 3	6.5	8.5
NAVCOMM 2	Narco Mk 12 360	11.0	13.1
VOR/ILS	Narco VOA 4	0.4	3.3
Add:			
DME	Narco DME 190	2.9	5.2
COM	King KY 197	6.0	3.2
NAV	Narco NAV 122	0.6	3.3
New totals		53.2	26.6

Now I was getting somewhere. I had saved almost 15 pounds and 6 amps. Let's keep it up . . .

Function	Type	Load (amps)	Weight (lbs)
Step Four			
Delete:			
ADF	Motorola ADF T 12B	1.3	10.9
	Davtron 701	0.6	0.4
Add:			
ADF	King KR 87	0.4	8.3
New totals		25.1	50.2

Now I was getting closer to what I had in mind. A couple more changes produced the package that I'm flying today:

Function	Type	Load (amps)	Weight (lbs)
Step Five:			
Delete:			
HSI	Narco DGO 10	1.5	4.7
Add:			
HSI	Narco HSI 100	1.6	4.4
VOR CVT	Narco OC 110		1.7
New totals		25.2	51.6

Up a bit in the weight and amperage departments. However, since I started, I have saved 16.3 pounds and 6.9 amps from my load

Fig. 7-1. The author's Comanche halfway through the upgrading process. So far, a Narco DGO 10 HSI and King MKA 20 audio panel had been added, along with those horrible eyebrow lights (courtesy Stu Leventhal).

(Fig. 7-2). Now I can transmit on the ground just using the battery—I don't have to start up. In the old days, when I had a generator (I replaced it with an alternator last year) and a Mark 12, I would have to start the engine and run it up to 1400 RPM to be able to transmit.

I haven't finished upgrading my package. I still intend to make a few changes. Here's the idea:

Function	Type	Load (amps)	Weight (lbs)
Step Six			
Delete:			
NAVCOMM 1	Narco Mk 12	11.0	13.0
VOR CVT	Narco OC 110		1.7
G/S	Narco UGR 2	0.2	3.1
DME	Narco DME 190	2.9	5.2
Add:			
NAV SYSTEM	King KNS 80	1.8	6.0
COM	King KY 92	4.5	2.8
New totals		17.4	37.4

When I get around to making these changes, I will save another 7.8 amps and another 14.2 pounds, for a total saving from the start of 15.3 amps and 30.6 pounds. These are significant improvements. To recap, my final package will look like this:

Function	Type	Load (amps)	Weight (lbs)
AUDIO	King KMA 20	1.1	2.3
ADF	King KR 87	0.4	8.3
COMM 1	King KY 197	0.6	3.2
COMM 2	King KY 92	4.5	2.8
NAV SYSTEM 1	King KNS 80	1.8	6.0
DVOR	Davtron 902A	0.0	0.4
HSI	Narco HSI 100	1.6	4.4
NAV 2	Narco NAV 122	0.6	3.3
XPDR	Collins TDR 950L	1.3	2.0
ENCODER	Aerosonic 3P	0.0	1.0
IVSI/GS DIR	Teledyne SLZ 9541	0.1	0.0
ELT	Narco ELT	0.0	3.7
Total		17.4	37.4

HOW TO JUSTIFY UPGRADING

Now, a lot of people say it's ridiculous to put all that money into an old airplane. "You'll never get your money out of it," they

Fig. 7-2. The author's Comanche at time of writing. A completely rebuilt panel, new 3-inch horizon, new Narco HSI 100S, NAV 122, KR 87 ADF, KY 197 COMM, Collins TDR 950 transponder and Narco DME 190 DME are already in place. To come: a KNS 80 in place of the Mark 12 and another KY197. The new KY 197 will go where the transponder is now, and the transponder will replace the DME 190. The DME 190 and Mark 12 will be sold. Note post lights have replaced the eyebrows. Panel by Linden Avionics.

say. It's true. But I'm not putting radios into the airplane to get my money out. I'm putting them in for maximum utility and reliability.

The way I look at it, my Comanche is a fine airplane—a classic. Its value will increase over the years. I like the plane and I have no intention of trading it in the near future. Why shouldn't I enjoy the airplane the way I want it? Flying with aging radios that keep breaking down is no fun. Flying it with *new* radios that keep breaking down is no fun either! Fortunately, when the new radios break down, it's usually a fast switch of a circuit board or other module, and all is well. For example, shortly after I put in the new King KR 87 ADF, it broke down. So we put in another one—brand new, under warranty. And *it* broke down. The problem was that the little digital readouts would flicker and then suddenly read all 8s, and we'd get no signal. When we put it on the bench, it took less than five minutes to isolate the problem as the computer circuit. Another circuit board was snapped in, and it has worked perfectly ever since. The reliability of the new radios is *much* greater than the old stuff.

Just remember to increase your insurance to reflect the increased value of your bird as you upgrade. Make sure the policy will pay a *stated value*, not a replacement value. Otherwise for sure you'll never get your money out.

Chapter 8

The Avionics Industry

Most avionics manufacturers *specialize* in avionics. The familiar home electronics names such as Zenith, Sony, General Electric, Pioneer, Panasonic, and so on, have not seen fit to get into this market, with the exception of RCA, which makes some components, but not a full line.

The avionics producers mostly address specific market segments, although a few attempt to serve the whole industry. The manufacturers that specialize in general aviation avionics are given in the table that follows.

Table 8-1 shows the manufacturer's name and address, whether their equipment is available for factory installation and/or retrofits, and the extent of their line—light, medium or heavy. These terms correspond roughly to the types of aircraft they would be installed in. If a maker tends to specialize in one or two particular areas, this is mentioned in a note.

THE MAKERS AND THEIR WARES

Here is a review of the current (1980) offerings of the principal avionics manufacturers.

The Bendix Corporation Avionics Division, Box 9414, Fort Lauderdale, Florida 33310. Bendix continues to offer its imaginative BX-2000 series, which is now available for factory installation on Mooney, Beech and some Pipers. The BX 2000 series (Fig. 8-1) features digital readout and optional keyboard tuning. All VHF frequencies are 200 channels for NAV and 720 for

COMM. All the components integrate with each other and are not particularly compatible with competitive brands. NAV displays may be on the revolutionary Bendix ECDI (electronic course deviation indicator) or through an HSI or RMI.

If you have a black and white digital Bendix weather radar, you can upgrade this to color radar for less than $6,400 (less a trade-in allowance). Bendix radios are 6.26 inches wide (Table 8-2).

Cessna Aircraft Company Aircraft Radio and Controls Division, Box 1521, Wichita, Kansas 67201. Censsna Avionics, which are the only kinds of radios offered in the Cessna aircraft line, are often sold in packages (Fig. 8-2). The basic package turns the airplane from a bare-bone model to a "II" designation. The "II" package varies with the aircraft. In the Skyhawk, for example, it consists of a 300 series NAVCOMM. A good 95 percent of the

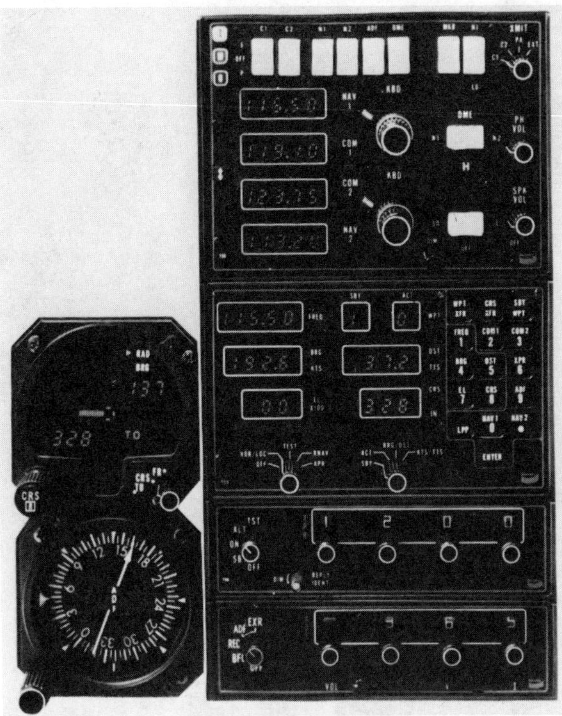

Fig. 8-1. The Bendix 2000 system. The top unit is the CN 2011A dual NAVCOMM and the audio panel, followed by the NP 2041A NAV computer programmer and keyboard for frequency selection, the TR 2016A transponder and the DF 2071A ADF. Note ECDI indicating aircraft is on the 137 degree radial and needs to steer left to get on the 148 degree radial inbound (courtesy Bendix).

Table 8-1. Avionics manufacturers.

Manufacturer & Address	Factory Installed	Retrofit Market	Avionics Line Light Medium Heavy
Aircraft Radio and Controls Division Cessna Aircraft Company Box 1521 Wichita KS 67201	Cessna only	No	Yes Yes Yes
The Bendix Corporation Avionics Division Box 9414 Fort Lauderdale FL 33310	Some	Yes	No Yes Yes
Bonzer, Inc. 90th and Cody Overland Park KS 66214	No	Yes	Yes Yes No
Brelonix, Inc. 106 N 36th Street Seattle WA 98103	No	Yes (HF only)	No Yes No
Collins General Aviation Division Rockwell International Cedar Rapids IA 52406	Yes	Yes	No Yes Yes
Davtron Inc. 427 Hillcrest Way Redwood City CA 94062	No	Yes	Yes Yes No
Edo-Air Group Marketing 216 Passaic Avenue Fairfield, NJ 07006	Some	Yes	Yes Yes No
Foster Airdata Systems, Inc. 7020 Huntley Road Columbus OH 43229	Some (RNAV only)	Yes	Yes Yes Yes

Company				
General Aviation Electronics, Inc. 4141 Kingman Drive Indianapolis IN 46226	No	Yes	Yes	No No
King Radio Corporation 400 N Rogers Road Olathe KS 66061	Yes	Yes	No	Yes Yes
Logue Avionics 337 Manchester Road Poughkeepsie NY 12603	No	Yes	No	Yes No
Mentor Radio Co. 1551 Lost Nation Road Willoughby OH 44094	No	Yes	Yes	No No
Narco Avionics Fort Washington PA 19034	Some	Yes	Yes	Yes No
RCA Avionics System Division 8500 Balboa avenue Van Nuys CA 91409	Some (Radar & DME)	Yes	No	Yes Yes
Ryan Stormscope 4800 Evanswood Drive Columbus OH 43229	No	Yes	No	Yes No
Sunair Electronics, Inc. 3101 SW 3rd Avenue Fort Lauderdale FL 33315	Some (HF only)	Yes	No	Yes Yes
Symbolic Displays, Inc. 1762 McGaw Avenue Irvine CA 92714	No	Yes	Yes	Yes Yes
Terra Corporation 3520 Pan American Freeway NE Albuquerque NM 87107	No	Yes	Yes	No No

Cessna light singles being sold are equipped with the II option, which includes other popular accessories as well. The next option is the NAV-PAC, which adds a second NAVCOMM, an ADF and a transponder in the Skyhawk series. The buyer saves about 18 per cent in buying the equipment this way.

Cessna's standard radios are 6.625 inches wide, vs 6.25 inches for King, Narco and Collins, meaning that they literally don't stack up well against competing brands. Cessna does not address the retrofit market at all. They make a full line of avionics, except for weather radar. Since you likely wouldn't be buying Cessna ARC radios for retrofit, I won't list them here, but, if your're buying a new Cessna, you'll want your salesman to give you the complete rundown.

Collins General Aviation Division Rockwell International, Cedar Rapids, Iowa, 52406. Collins continues to produce its Micro Line radios, which come either with digital frequency presentation, giving storage of the next-required frequency, or (at lower cost) with conventional mechanical frequency presentation. Collins also offers a new HF radio, the HF-200, which features a frequency synthesizer to provide any 20 channels in the 2 to 22.9 MHz range, in 100 kHz steps. Frequencies are programmable on the ground. Collins radios stack to a 6.25 inch width (Fig. 8-3).

Collins produced the DCE-400 (Fig. 8-4), a unique device that could be described as a "poor-man's DME." It did not sell too well, however, even though the concept is sound. It is not a radio, but a

Fig. 8-2. Cessna avionics are found almost exclusively in Cessnas. Here's a 1980 Cessna 180 with a full load (courtesy Cessna).

Table 8-2. Bendix avionics.

Item	Model	1980 Cost	Remarks
AUDIO	AS 2015A	$350	Not needed with CN 2011A
NAVCOMM	CN 2011A	$7,213	Dual NAVCOMM and audio panel Price includes ECDIs
NAVCOMM	CN 2012A	$3,873	Single NAVCOMM COMM frequency preselect Price includes ECDI
NAVCOMM	CN 2013A	$3,028	Single NAVCOMM Price includes ECDI
CDI	IN 2014A	$1,100+	Electronic CDI (ECDI) Various options available
HSI	HSD 880	$5,665+	Price depends on options Includes slaved gyro system
DME	DME 2030	$3,995	DME can readout on ECDI
RNAV	NP 2041A	$5,004	Keyboard can tune BX 2000 NAVCOMMs Entries may be made with programmable pocket calculator
ADF	DF 2071A	$1,595	RMI available
XPDR	TR 2016A	$720	
RADAR	RDR 160	$7,967	BW
RADAR	RDR 160	$11,285	Color Optional checklists may be displayed NAV may be displayed from BX 2000 system

computer. You hook it up to your dual VORs and make settings to it and it gives you distance along track and ground speed. Although it gives DME-like information, it is not legal for use where a DME is required. However, at less than $1,200 it is considerably cheaper than a DME, and it could well be the answer to getting more information out of your dual VORs at relatively low cost (Table 8-3).

Edo-Aire Group Marketing 216 Passaic Avenue, Fairfield, New Jersey 07006. Edo-Aire produces a series of self-contained one and one half NAVCOMMs, complete with VOR indicator and either 360 or 720 channel COMMs, all in one box (Fig. 8-5). Optional "automatic omni" makes the needle center with a "to" reading when the OBS knob is pushed, giving an immediate bearing to the station. A cheaper version, which shares the receiver, giving either NAV or COMM, but not both simultaneously, is the RT 553 series. Single COMMs and NAVs are also available. Edo-Aire radios are 6.5 inches wide, meaning they are ¼ inch wider than Collins, King and Narco units.

Edo-Aire also offers an ADF with a built in approach timer and indicator, and a DME, the RT 888. This features a simultaneous digital distance and groundspeed readout, with time-to-station and

Fig. 8-3. Collins VIR 351 COMM features frequency storage and electronic digital readout (courtesy Collins).

Item	Model	1980 Cost	Remarks
AUDIO	AMR 350		$635 Includes marker
COMM	VHF 250	$1,2757	20 channels Mechanical frequency readout
COMM	VHF 251	$1,595	720 channels Electronic frequency readout and storage
NAV	VIR 350	$1,740	Mechanical frequency readout Price includes CDI
NAV	VIR 351	$2,075	Electronic frequency and to/from bearing readout Price includes CDI
CDI	IND 350A	$445	VOR/LOC indicator
CDI	IND 351A		$580 VOR/LOC/GS indicator
HSI	PN 101	$7,150	Includes slaved compass system
DME	DME 451	$3,980	Includes clock, groundspeed and time-to-station
DCE	DCE 400	$1,180	Distance computing equipment Requires dual VORs
RNAV	ANS 351	$3,1208	waypoints Groundspeed to waypoint
ADF	ADF 650	$1,895	
XPDR	TDR 950	$720	
RADAR	WXR 150	$7,240	BW digital display
HF COMM	HF 200	$7,490	20 programmable channels

timing functions being available instead of groundspeed when required. Edo-Aire's "Squareponder" RT 667 is the only transponder currently available that will fit in a standard 3″ instrument hole. A flat pack version is also available. Edo produces both slaved and

Fig. 8-4. Collins DCE 400, the "poor man's DME" (courtesy Collins.).

unslaved HSIs, as well as autopilots and a flight director system (Table 8-4).

General Aviation Electronics, Inc. (Genave) 4141 Kingsman Drive, Indianapolis Indiana, 46226. Genave offers its new GA/1000 NAVCOMM, which is a complete one and a half system featuring digital frequency readout, 720 COMM channels, 200 NAV channels and the option of locating the VOR indicator either within the panel-mounted radio or elsewhere on the instrument panel, as desired (Fig. 8-6). It costs $1,395 (plus installation). Genave also brought out its Alpha 720 COMM transceiver, priced at $1,045. This has a unique feature—its own speaker located right in the front panel. The new Alpha Six COMM is a six channel hand-held unit resembling a walkie-talkie, weighing 2.25 lbs with its own self-contained battery. This can be

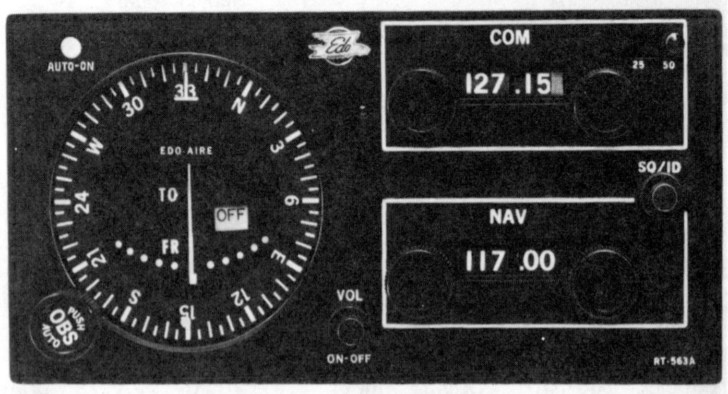

Fig. 8-5. Edo-Aire 563A 720-channel COMM and 200-channel VNA, combined with CDI and automatic omni feature, all in one box (courtesy Edo-Aire).

Table 8-4. Edo-Aire avionics.

Item	Model	1980 Cost	Remarks
AUDIO	AM 550	$550	Includes marker beacon Also available TSO'd (AM 660) and without marker (A 550)
NAVCOMM	RT 553	$1,395	NAVCOMM with shared receiver Built-in CDI 360 channels 720 channels available (RT 553A) Automatic omni available
NAVCOMM	RT 563	$1,895	NAVCOMM with simultaneous dual receivers for NAV and COMM 360 channels 720 channels available (RT 563A) Automatic omni standard
COMM	RT 551A	$1.095	720 channels Also available TSO'd (RT 661A) and with 360 channels (RT 551)
NAV	R 552	$1,430	Price includes CDI Also available TSO'd (R 662)
NAV	R 554	$2,168	Includes GS receiver Price includes CDI Also available TSO'd (R 664)
CDI	CID 552A	$675	VOR/LOC only Also available TSO'd (CID 662)
CDI	CID 554A	$750	Includes GS needle Also Available TSO'd (CID 664)
HSI	NSD 360A	$2,783	Unslaved
HSI	NSD 360A(S)	$4,182	Slaved
DME	DME 888	$3,995	Remote tuning from NAV receiver
ADF	R 556E	$1,425	Digital frequency presentation Analog tuning
XPDR	RT 667	$845	Fits iin 3" instrument hole
XPDR	RT 777	$745	Flat pack
XPDR	RT 887	$995	Fits in 3" instrument hole Incandescent digital display

Fig. 8-6. Genave GA/1000 NAVCOMM features 720-channel COMM, 200-channel NAV and VOR/LOC CDI, all in one box (courtesy Genave).

factory set for any six channels in the 720 channel VHF range. It has a two watt transmitter, and an integral speaker/mike. The unit has been popular with pilots as an emergency IFR stand-by unit. It sells for only $495.

Genave radios are 6.5 inches wide, meaning they are ¼ inch wider than Collins, King and Narco (Table 8-5).

King Radio Corporation 400 N Rogers Road, Olathe, Kansas, 66061. King offers two levels of radios in its Silver Crown line—the older ones with mechanical frequency readouts and frequency storage. It also offers the Gold Crown line for airline and corporate aircraft, not discussed here. King has been doing very well with its unique KNS 80 VOR/DNE/GS/RNAV, and has recently introduced a smaller, 9-waypoint version, the KNS 81

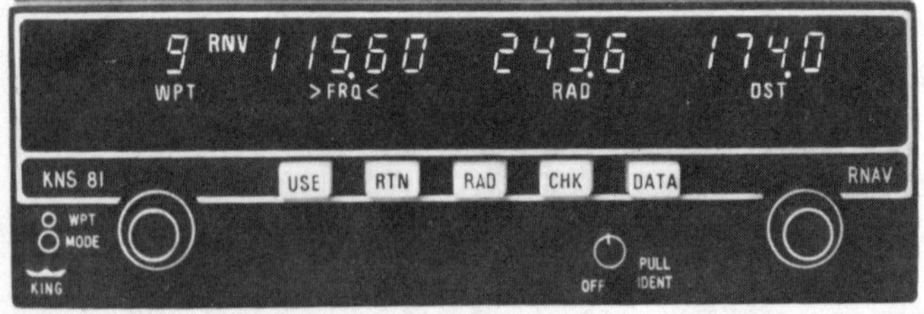

KNS 81 INTEGRATED NAVIGATION SYSTEM (VOR / ILS / RNAV)
(SHOWN IN RNAV MODE)

Fig. 8-7. New King KNS 81 9-waypoint RNAV system (Courtesy King).

Item	Model	1980 Cost	Remarks
AUDION	Tau/200	$295	Marker readout optional Includes amplifier
MARKER	Delta/303	$139	Can readout through audio control head
NAVCOMM	Alpha/200B	$1,100	100 channel COMM 100 channel NAV
NAVCOMM	GA 1000	$1,395	720 channel COMM 200 channel NAV Includes CDI Digital frequency
COMM	Alpha/720	$1,045	720 channel COMM Built-in speaker Digital frequency
ADF	Sigma/1500	$1,045	Crystal controlled
XPDR	Beta/5000	$650	

(Fig. 8-7), that needs a separate King DME. The new digital series, including the KY 196/197 720 channel COMM, the KN 53 200 channel NAV, the KN 62A DME and the KR 87 ADF are fast becoming popular additions to pilot's panels (I have the KY 197 COMM and KR 87 ADF in my Comanche). King also has a new audio-control panel, the KMA 24. In addition, the new KN 63 DME (Fig. 8-8) has been announced, which is a remote unit with a panel readout giving simultaneous distance, groundspeed and time-to-station. This is available with a duplicated readout for the copilot, if desired.

King also offers a flight director, autopilot and weather radar. King radios are 6.25 inches wide (Table 8-6).

Narco Avionics Fort Washington, Pennsylvania, 19034. Narco Avionics has recently been aquired by Edward Zimmer of Berkely Industries, an electronics manufacturer. The company is concentrating on its popular Centerline products (Fig. 8-9), with great promise being held out as a result of its new ownership.

In the painful process of reevaluating its marketing philosophy, Narco designed, introduced, and then withdrew from the marketplace an exotic line of radios called the "E-Line." These featured electronic frequency readout, frequency storage and other

Table 8-6. King avionics.

Item	Model	1980 Cost	Remarks
AUDIO	KMA 24	$675	Includes amplifiers for both speaker and phones Includes marker beacon
AUDIO	KA 134	$310	Switcher and amplifier only
NAVCOMM	KX 175B	$1,990	Mechanical frequencies 200 NAV/720 COMM channels Available non-TSO'd as KX 170B
NAVCOMM	KX 145	$1,210	Mechanical frequencies 200 NAV/720 COMM channels Shared receiver between COMM and NAV
COMM	KY 92	$1,150	Mechanical frequencies 720 channels
COMM	KY 196 KY 197	$1,755 $1,755	Electronic frequencies Frequency preselect and storage 720 channels KY 196 is 28 volt version KY 197 is 14 volt version
NAV	KN 53	$1,865	Electronic frequencies Frequency preselect and storage 200 channels
CDI	KI 203	$880	VOR/LOC with built-in converter Available with GS needle as KI 203
CDI	KI 206	$720	VOR/LOC/GS without converter
CDI	KI 208	$555	VOR/LOC with built-in converter Available with GS needle as KI 209

Type	Model	Price	Notes
HSI	KCS 55A	$3,515	Includes slaved compass system
RMI	KI 226	$2,205	Needs KCS 55A system to run it
DME	DN 62A	$3,100	TSO'd DME Maybe remotely channelled
DME	KN 63	$3,850	Remote DME
XPDR	KT 76A	$730	Also available for use below 15,000 feet as KT 78A
ENCDR	KE 127	$835	Remote encoding altimeter
RNAV	KNS 80	$5,950	Self-contained 200 channel NAV, DME, GS, converter and 4-waypoint RNAV Available without GS
RNAV	KNS 81	$4,050	Self-contained 200 channel NAV, GS, converter and 9-waypoint RNAV Available without GS
ADF	KR 86	$1,495	Crystal controlled ADF with built-in indicator
ADF	KR 87	$1,925	Electronic frequencies Frequency preselect and storage Built-in timers
RADAR	KWX 50	$6,795	BW Digital radar
RADAR ALT	KRA 10	$2,200	20 to 2,500 feet
FLT DIR	KFC	$10,450	Includes KCS 55A HSI and slaved compass system

Fig. 8-8. This KN 63 remote unit goes with KNS 81 in Fig. 8-7 (courtesy King).

modern innovations. They did the same thing with a 10-waypoint RNAV—the RNAV 161.

Narco has a couple of unique products—the NAV 121 and NAV 122, which give a complete NAV system, including tuner, in one box. (I have a NAV 122 in my Comanche, and I am very pleased with it. I also have a Narco HSI 100S slaved horizontal situation indicator and a DME 190, which is the best-selling DME in the country.) Narco radios are 6.25 inches wide (Table 8-7).

Fig. 8-9. The Narco Centerline features, from left to right and top to bottom, NAV 121, NAV 122, CP 136, ADF 141, COM 120, AT 150, ADF indicator, HSI 100, DME 195, DME 90 and NAV 124 (courtesy Narco).

OTHER AVIONICS WARES

As mentioned, there are several manufacturers that specialize in one particular area of aviation radio:

Bonzer, Inc. 90th and Cody, Overland Park, Kansas, 66214. Bonzer produces radar altimeters, including the lowest cost Mini-Mark ($995). This features an optional decision height annunciator. They also make the Mark 10X, which costs about $2,500.

Brelonix, Inc. 106 N 36th Street Seattle, Washington, 98103. Brelonix specializes in HF COMMs. Their lowest price unit is the SAM 70, a five-channel unit at $1,595. They also make the SAM 100/5 and 100/10, five and ten channel units priced at $2,250 and $2,650 respectively.

Table 8-7. Narco avionics.

Item	Model	1980 Cost	Remarks
AUDIO	CP 136	$435	Push-button selection
COMM	COM 120	$1,345	720 channel COM
NAV	NAV 121	$1,250	Self-contained VOR/LOC indicator and tuner in one 3″ hole
NAV	NAV 122	$1,975	As NAV 122, plus GS and marker receivers and indicators
NAV	NAV 124	$1,855	200 channel NAV receiver tod drive HSI Includes VOR converter
HSI	HSI 100	$2,680	Unslaved HSI
HSI	HSI 100S	$3,495	Slaved HSI
DME	DME 190	$2,750	DME, groundspeed and time-to-station, with tuner Can be remotely tuned
DME	DME 195	$3,925	Remote version of DME 190
XPDR	AT 150	$665	
ENCDR	AR 500	$725	Blind encoder

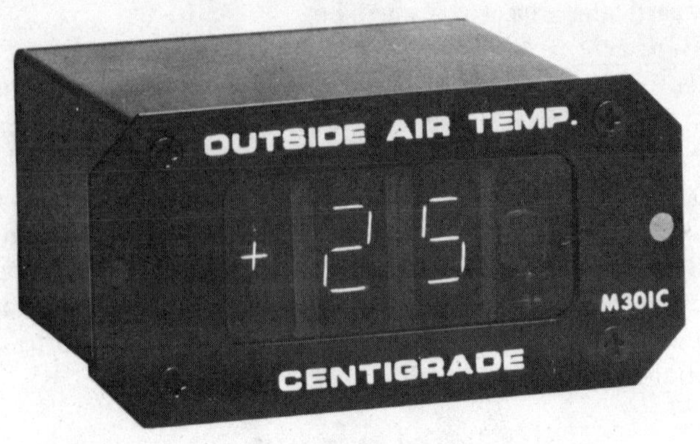

Fig. 8-10 Davtron digital outside air temperature gauge for the pilot who must have everything (courtesy Davtron).

Davtron Inc. 427 Hillcrest Way Redwood City, California, 94062. Davtron produces a variety of digital readout devices. One, the 902A, gives digital VOR RMI data. It sells for under $300. Another, the 701A, shows the ADF requency you have tuned with an analog tuner (under $200). They also have an intercom, an outside air temperature gauge (Fig. 8-10) and a couple of clocks and timers (Fig. 8-11).

Foster Airdata Systems, Inc. 7020 Huntley Road Columbus, Ohio, 43229. Foster (no relation to the author) produces RNAV systems. Their 511 is a 2-waypoint RNAV that works with virtually all NAVs and DMEs. It presents its steering information digitally, and an optional needle steering converter is available to make a CDI or HSI do the work (about $1,850+). The 612 is a 4-waypoint system that sells for about $5,000. Foster also has a digital VNAV (Fig. 8-12) which works with any of its RNAVs, and a digital ADF frequency indicator.

Logue Avionics 337 Manchester Road Poughkeepsie, New York, 12603. Logue recently introduced the LA 7800 electronic CDI, which works with most NAV receivers. It provides digital RMI information and costs around $1,000.

Mentor Radio Co. 1551 Lost Nation Road Willoughby, Ohio, 44094. Mentor sells low cost COMM and NAV units to the retrofit

Fig. 8-11. Davtron digital clock features countdown timer as well (courtesy Davtron).

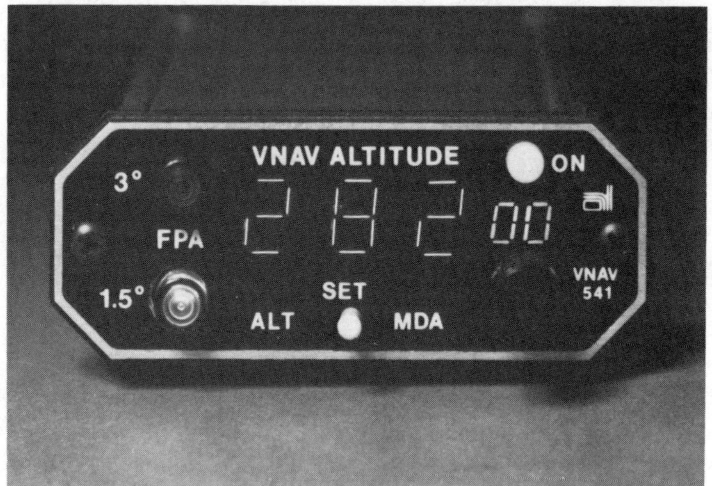

Fig. 8-12. Foster Airdata VNAV 541 works with any Foster RNAV to give precise descent command altitudes for 1.3 and 3 degree descents (courtesy Foster Airdata).

Fig. 8-13. RCA Weatherscout II radar in a Piper Lance. The other avionics are by King (courtesy RCA).

and homebuilt markets. The M 360 COMM sells for just under $1,000. It has 360 channels, and is available in a portable model. The M 200 NAV receiver works with a Mentor VL 2 CDI. The pair costs about $1,300 complete.

Fig. 8-14. RCA antenna fits in the leading edge of the wing on a Piper Lance (courtesy RCA).

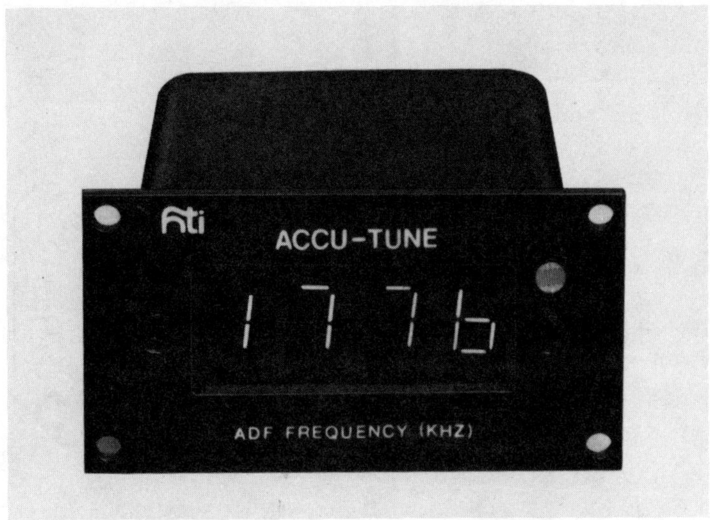

Fig. 8-15. Symbolic Displays Accut-Tune ADF digitizer (courtesy SDI).

RCA Avionics Systems Division 8500 Balboa Avenue Van Nuys, California, 91409. RCA has moved into the general aviation field with its new Weatherscout radar (Fig. 8-13), which can be installed in single engine aircraft without the need for a wing pod. The antenna is so flat it can fit in the wing leading edge (Fig. 8-14). This is now offered as a Piper factory option on the Saratoga aircraft. This radar sells for about $5,500.

Ryan Stormscope. 4800 Evanswood Drive, Columbus, Ohio, 43229. Ryan offers one product, the Stormscope. This unique FAA-approved weather avoidance system sells for under $6,000, and is fast becoming a best-seller.

Sunair Electronics, Inc. 3101 SW 3rd Avenue Fort Lauderdale, Florida, 33315. Sunair specializes in HF COMMs, and offers several models raging in price from under $3,000 to almost $11,000.

Symbolic Displays, Inc. 1762 McGaw Avenue Irvine, California, 92714. SDI produces a variety of digital VOR RMIs and digital ADF frequency readouts (Fig. 8-15), as well as fuel computers and other electronic devices.

Terra Corporation 3520 Pan American Freeway NE Albuquerque, New Mexico, 87107. Terra sells the Radair line of low-cost COMMs and NAVs to the retrofit and homebuilt market. The R 360 COMM sells for about $575, while the R200 NAV receiver is only $395, plus $350 for the CDI-converter.

Chapter 9
Getting On With It

It's all very well to read and study and dream and plan and scheme. But what about *getting on with it*? Here's a summary of the action steps.

1. Evaluate your avionics needs for the type of flying you do or are going to do. This procedure is outlined starting in Chapter 4.

2. Find out what's wrong with your current avionics. This is covered in Chapter 5.

3. Select the radios you want. This is covered in Chapter 6. If you weren't sure about what's available, Chapter 8 reviewed the current marketplace.

4. Get on with it! That's the subject of this chapter.

HOW BIG A DEAL?

Upgrading may be simply a matter of taking out one radio and replacing it with another. Or it may call for a complete panel rebuild. Assuming that your desires are a bit more complex than a simple remove-and-replace job, how can you plan for the big fix? A good solution is provided by the major manufacturers. Collins, King and Narco all publish large poster-like sheets depicting scale drawings of their radios. Many of the makers' brochures on individual radios include a full-scale illustration. Get a large sheet of bristol board and lay out your panel on it, to scale. Then, using the scale illustrations from the brochures or posters, play around with how you want to lay things out. Narco Avionics used to offer extremely useful scale instrument panel planning sheets for most

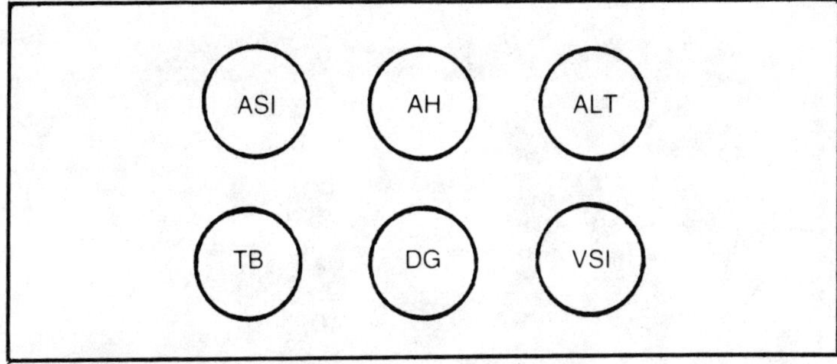

Fig. 9-1. Standard "T" format panel layout.

light aircraft, along with a set of sticky cutouts of their avionics, all for $5. They may still have one for your airplane left. It's worth a try. (Write to Narco at Fort Washington, Pennsylvania, 19034).

If you are redesigning your instrument panel, try to set it up so that the instruments are in the standard "T" format (Fig. 9-1).

A lot of panels in older aircraft must have been laid out by the simple expedient of applying some glue to the basic panel and throwing the instruments at it. Wherever they landed was where they went. The standard "T" layout has since been established as the best way to distribute instruments for easiest IFR flying.

Going from the basic "T," Figure 9-2 shows my suggestions for the disposition of avionics.

Try to avoid putting the ADF dial over on the right side of the panel (Fig 9-3). That's the wrong place if you want to do an ADF approach with any comfort. If your ADF tuner includes the dial, put

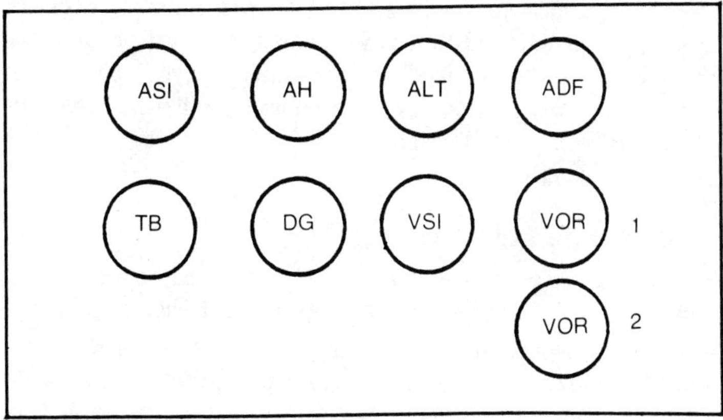

Fig. 9-2. Recommended layout of avionics.

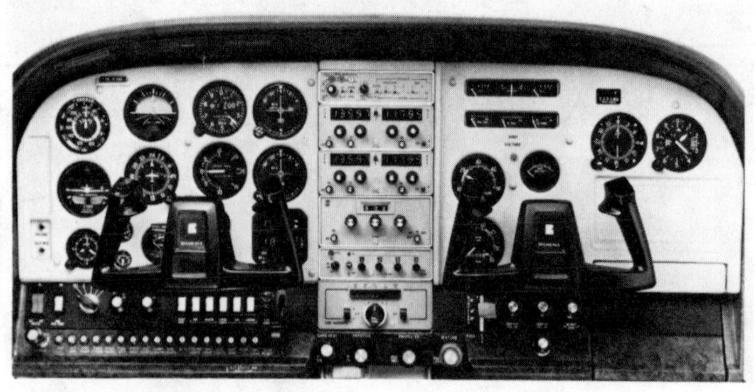

Fig. 9-3. The placement of the ADF dial in this Cessna Skylane RG, on the far right, is unacceptable. The problem could be solved by installing an HSI where the DG is and putting the ADF dial where the top VOR dial is, then putting the DG where the ADF dial is now, as a standby instrument (courtesy Cessna).

the ADF at the top of the center radio stack if possible, so that the dial will be within your scan (Fig. 9-4).

Try to get the appropriate tuner to line up with the relevant CDI or dial, as in Figure 9-5.

I have been harping about the benefits of an HSI. One of the most significant is that it saves you an instrument hole, since it combines your DG and VOR in one instrument. If you have an HSI, it should go in the center of the stack, beneath the horizon, and it should ideally be tuned off NAV 1 (Fig. 9-6).

If you install an HSI, and you have room to spare, save your old DG and install it somewhere as a standby unit. If the old DG is air driven, get an electric HSI, and vice-versa. This way you'll have a redundant system for heading information. If your vacuum or pressure pump fails (they do, you know), you'll still have your electric HSI. If your alternator or generator fails, you'll still have your air-driven DG.

LAYING OUT THE RADIO STACK

The ideal placement for your radios is in a center stack. However, with an older aircraft, you may not have this luxury. Older Comanches, Apaches, Bonanzas, and Cessnas used to put radios in glove boxes—some on the extreme left, some on the extreme right. If this is your lot, you may try to tilt the right hand radios toward you a bit, to make them easier to see. Some of these

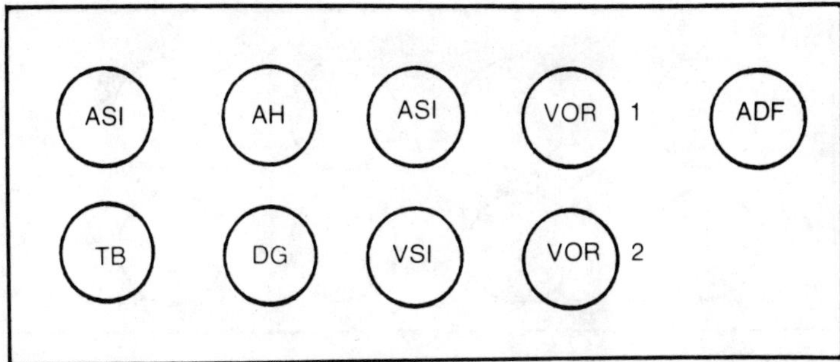

Fig. 9-4. Better placement of ADF dial.

can be converted. I know that many owners of older Comanches have gone to a center stack arrangement (Fig. 9-7).

If you have a center stack, a suggested layout is shown in Fig. 9-8.

WHAT ABOUT RADAR?

If you are retrofitting radar, try to get the screen up where you can see it. The example in Fig. 9-9 is a poor location, since it is a dangerous procedure to have to keep looking down when you're flying on instruments. To do so can cause vertigo, since the sharp and repetitive head movements you'll need to keep your eye on the radar will screw up your inner-ear equilibrium and make you think you're doing something you're not.

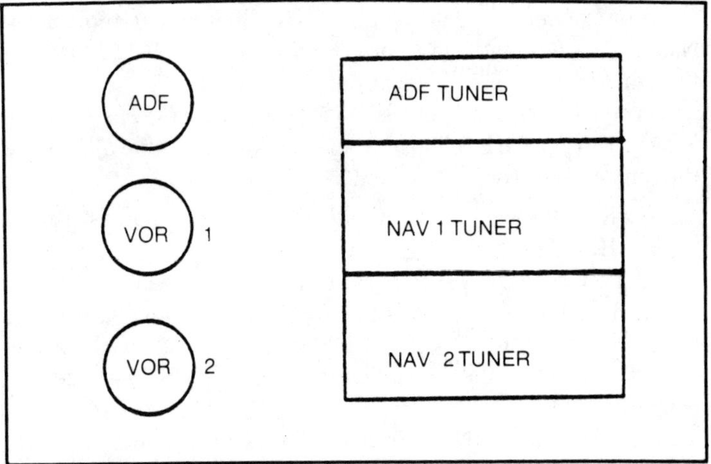

Fig. 9-5. Coordinated placement of tuners and dials.

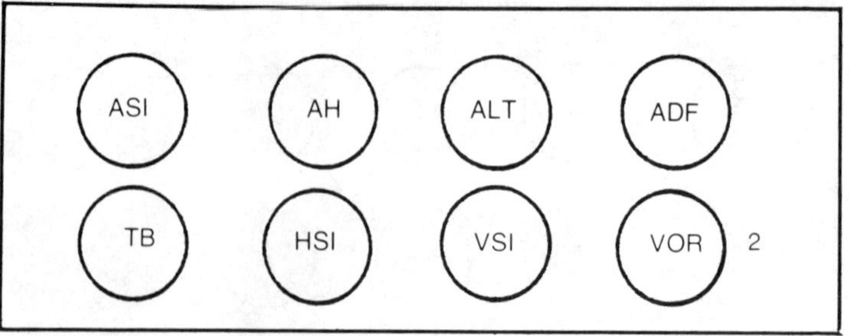

Fig. 9-6. Good placement of HSI.

THE IMPORTANCE OF GOOD PLANNING

I can't stress too highly the need to plan carefully when upgrading your avionics. Ripping things out and putting them back in is a time-consuming and expensive process. If you can afford it at all, try to do everything at once, so you have only one big tear-down. To do otherwise can run up your installation bills, and add to the wear and tear on your existing components. There's always the chance that a wire will get broken if there's a lot of activity going on behind the panel every two months. However, if you can only afford to do one or two things at a time, which was my situation, think very carefully about the sequence in which you are going to do things. Make the most critical fixes first. Replace non-functioning NAVs and COMMs before going to RNAV and radar. Give much consideration to the consequences of additions and deletions.

For example, I want to put in a KNS 80 RNAV to replace my Narco Mark 12 in the radio stack. Figures 9-10 and 9-11 show how the panel of my Comanche looks before the move.

The KNS 80 is the same size as the Narco Mark 12, so the plan is simply to put the KNS 80 in where the Mark 12 is now. Here are the consequences of this decision:

1. KNS 80 will not work with DGO 10 HSI, so first I must replace HSI, therefore...

2. I replaced DGO 10 with Narco HSI 100s, however...

3. HSI 100s needs a VOR converter, so I arranged with my radio shop to rent me a converter on a temporary basis on the understanding that they would later sell me a KNS 80 and get the converter back.

4. When the KNS 80 goes in and the Mark 12 comes out, I will be shy my stand-by COMM, so I must get another COMM. Where will this go, since my stack is full? Well...

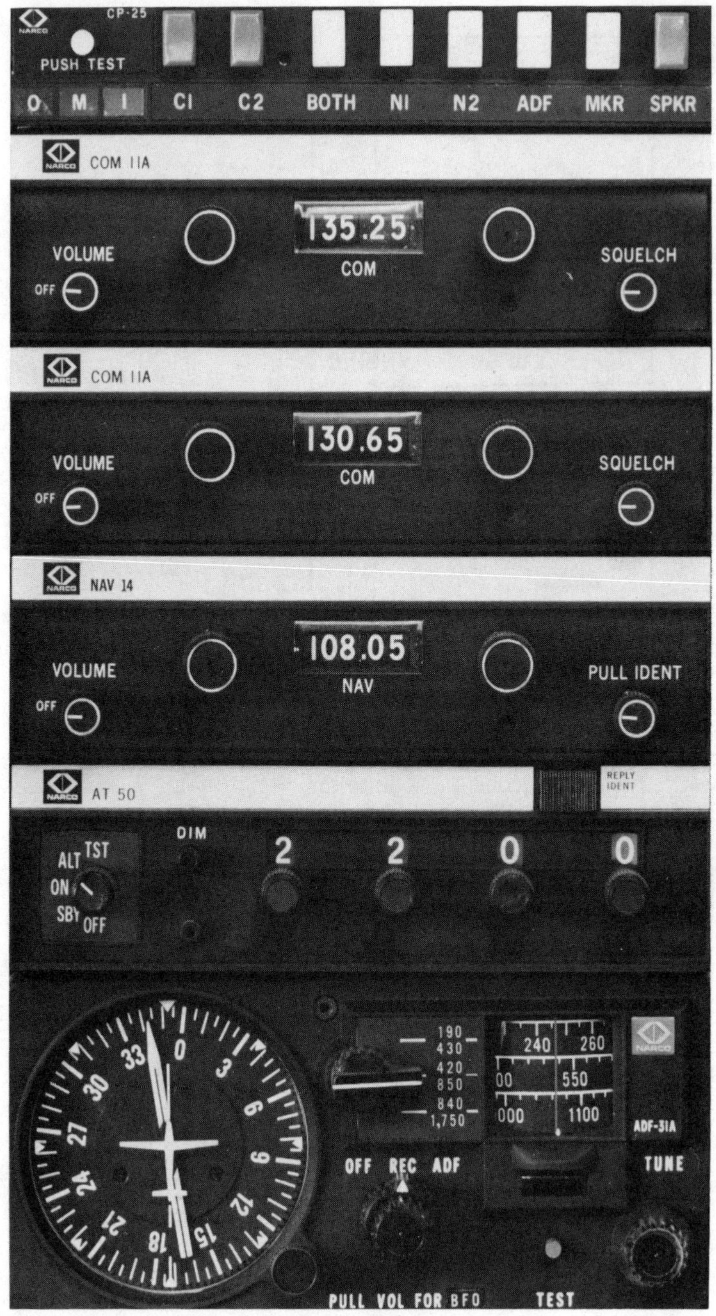

Fig. 9-7. Aircraft radio center stack (courtesy Narco).

AUDIO PANEL
ADF TUNER
NAVCOMM 1
NAV COMM 2
DME
TRANSPONDER

Fig. 9-8. Suggested layout of radio center stack.

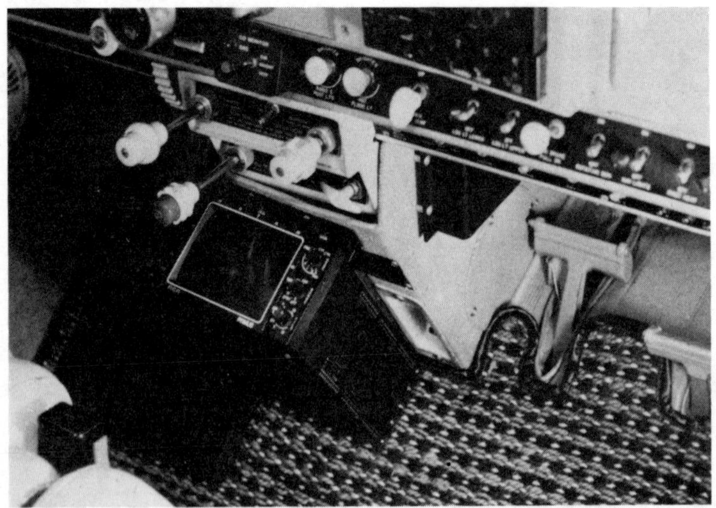

Fig. 9-9. This is a poor place to locate the radar (courtesy RCA).

Fig. 9-10. The author's Comanche "almost there."

5. When the KNS 80 goes in, I'll no longer need my DME 190, since the KNS 80 includes its own DME, so I'll sell the DME 190, thus reducing my KNS 80 cost by about $1,500. However, this is about the cost of the second COMM I'll need. I could put the new COMM where the DME 190 is now, but that would get the second COMM out of the center stack, which I don't want . . .

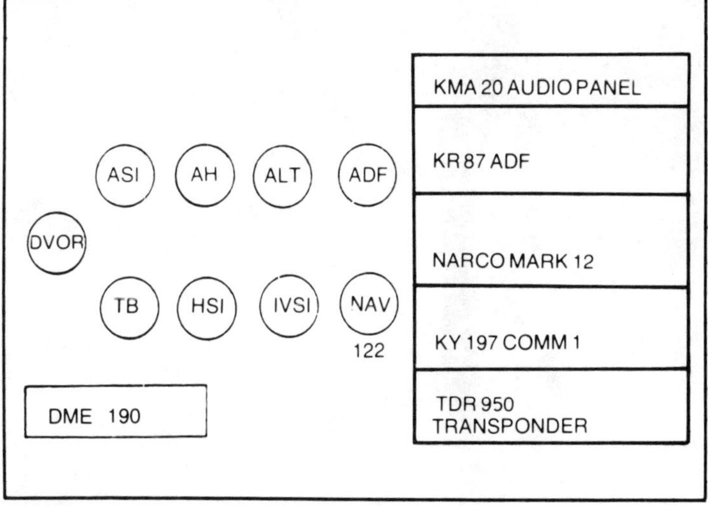

Fig. 9-11. Layout of instrument panel in author's Comanche.

6. So I'll put the new COMM where the transponder is now and move the transponder to where the DME is now.

7. When the KNS 80 goes in, I'll no longer need my UGR-2 glideslope receiver, since the KNS 80 includes that, so I'll sell the old GS . . .

8. And I'll sell the old Mark 12.

Here are the financials:

	Proceeds	*Expenses*
1. Sell DGO 10	$500	
2. Buy HSI 100S		$2,000 (used)
3. Install HSI and remote compass		$500
4. Rent Converter		$50
5. Buy KNS 80		$6,000
6. Sell Mark 12	$300	
7. Sell UGR 2	$100	
8. Buy COMM 2		$1,200
9. Sell DME 190	$1,500	
TOTAL	$2,400	$9,750
Net cost		$7,350

Chapter 10
Avionics
Troubleshooting

When your avionics go wrong, it can be very frustrating. Inevitably, when you take the offending radio out and put it on the bench, it works fine. Or fixing one radio results in another one going out. Here are some reasons for avionics problems:

☐ You don't understand how to use the radio and you just think there's something wrong (very common).

☐ There's something wrong with the radio.

☐ There's nothing wrong with the radio, but there is a problem in the installation (loose wire, faulty connection, shorted wire, vibrating mount, etc.)

☐ There's nothing wrong with the radio and the installation, but there's something wrong with the antenna (poor location, broken lead, water damage, etc.)

☐ There's nothing wrong with the radio, but its proximity to another radio causes an incompatibility problem (poor shielding, poor splitter connection, etc.)

☐ There's nothing wrong with the radio, but the station you're trying it on is off the air or being played with by the technicians.

A PROCESS OF ELIMINATION

Avionics troubleshooting is a process of elimination. The idea is to isolate all the good things until you finally get down to what is causing the problem. If you have a problem, make up a sheet as

Table 10-1. Avionics troubleshooting checklist.

Aircraft type   Registration

List radio package:

Which radio is causing problem?
Describe problem:

Has this problem happened before? () No () Yes
Was it fixed before? () No () Yes
What did they do?

Is problem () permanent, () intermittent?
When was equipment installed?
Have you had any other avionics troubles recently? () No
() Yes

If there is second system, does it have same problem? () No () Yes
When this problem occurs does everything else work? () No () Yes
Have you checked all circuit breakers and fuses? () No () Yes
Is battery fully charged? () No () Yes
Is generator or alternator system working properly? () No () Yes

shown (Table 10-1) and answer all questions. Then show the sheet to your friendly radio man. If he doesn't immediately say, "Eureka! *Here's* your problem!" invite him up for a ride and show him in the air. Leave the sheet and the radio (and the airplane, if necessary) with the radio shop, and check back regularly to see how they're doing. If the radio giving problems is old and almost worthless, ask if it's worth fixing. It may be better to bite the bullet and *upgrade*!

WHERE TO GET MORE INFORMATION

There are two good books that will help you understand more about the insides of your avionics. One is *Every Pilot's Guide to Aviation Electronics,* by John Ferrara, available at $9.75 from El Jac Publishing Company, 133A Glendale Avenue, Trenton, NJ 08618; the other is *Aviation Electronics Handbook*, by Edward L. Safford, available from TAB BOOKS Inc. Blue Ridge Summit, Pennsylvania, 17214.

In addition, I have found *The Aviation Consumer* to be very helpful in its unbiased reporting on avionics problems and solutions. This magazine comes out twice a month, and may be ordered at 1111 East Putnam Avenue, Riverside, Connecticut 06878 ($39 a year).

98

Trade-a-Plane carries the most comprehensive selection of used radio offerings. Order from them at Crossville, Tennessee, 38555 ($9.50 a year).

For comparative information of the latest avionics, get a copy of *Flying Annual* (published each February), *Business and Commercial Aviation Planning and Purchasing Handbook* (published each April) or *AOPA Pilot* (June issue covers avionics).

Appendix A
Panel Portfolio

Fig. A-1. This Commanche 400 has a King KCS 55A HSI, Collins ADF and RMI, Ryan Stormscope, KMA 20 audio panel, KNS 80 RNAV, KY 196 COMM, KX 175 NAVCOMM and an old Narco transponder. It also has an IVSI and a Century III autopilot with couplers.

Fig. A-2. Here's a Twin Comanche in which the owner built a wood veneer overlay. Narco COMM 111Bs, Narco audio panel and transponder surround a King KNS 80. Also note Narco NAV 12 and old Motorola ADF; calculator and stopwatch mounted on wheel.

Fig. A-3. This Comanche 400, G-POWA, was flown over the Atlantic from England with one Narco Mark 12. It returned with this superb, totally customized package of full King Silver Crown avionics, including KCS 55A HSI, radar altimeter, Ryan Stormscope and Century III autopilot.

Fig. A-4. Another Comanche 400. This one features a customized, etched lucite panel, with all markings engraved and edge lit, so that at night all the instruments and markings light up beautifully. This package includes the Narco Centerline, an old Cessna ARC ADF and the original Altimatic autopilot. HSI is a Narco DGO 10. Note Davtron digital VOR to the right of the HSI, buried in the panel.

Fig. A-5. The author finally got it together and installed a KNS 80. Now he plans to add another KY 197 and replace the KMA 20 with a KMA 24, both fitting at the top, above the KR 87 ADF. Also, a Ryan Stormscope would be nice, in the hole to the right of the ADF indicator, and a standby DG below that, being part of the intended Century II autopilot that is planned.

Appendix B
Principal Avionics Components Sold in the Last 20 Years

This appendix contains lists of the principal avionics components sold in the United States in the last twenty years or so. The lists are organized by type of component, e.g., COMM, Transponder, ADF, etc. They show the maker, the model number and whether the equipment is in current production (1980).

VHF Communications Transceivers

Maker	Model	Channels	Current (1980)
Bayside	BEI 990	90	No
Bayside	BEI 1060	360	No
Bendix	RT 241B	720	No
Bendix	RTA 41	360	No
Bendix	RTA 43A	720	No
Cessna ARC	RT 302	360	No
Cessna ARC	RF 432A	360	No
Cessna ARC	RT 432A	360	No
Cessna ARC	RT 524A	360	No
Cessna	1038A	720	Yes
Collins	618F	360	No
Collins	618M	720	No
Collins	VHF 250	720	Yes
Collins	VHF 251	720	Yes
Dynair	SKY 515A	360	No

VHF Communications Transceivers

Maker	Model	Channels	Current (1980)
Edo-Aire	RT 551	360	Yes
Edo-Aire	RT 551A	720	Yes
Edo-Aire	RT 661	360	Yes
Edo-Aire	RT 661A	720	Yes
Edo-Aire	RT 773	360	No
Edo-Aire	RT 771A	360	No
Edo-Aire	RTD 771	360	No
Genave	Alpha/100	100	No
Genave	Alpha/100	360	No
Genave	Alpha/720	720	Yes
King	KY 90	90	No
King	KY 92	720	Yes
King	KY 95	360	No
King	KY 195B	720	Yes
King	KY 196	720	Yes
King	KY 197	720	Yes
Mentor	M 360	360	Yes
Narco	Mark 7	360	No
Narco	COM 10	360	No
Narco	COM 11A	360	No
Narco	COM 11B	720	No
Narco	COM 111	360	No
Narco	COM 111B	720	No
Narco	COM 120	720	Yes
Narco	COM 120-20	720	Yes
Radair	360	360	No
Radair	R 360	360	Yes
Radair	SKY 515A	360	No
Skycrafters	SKY 515	360	No
Skycrahfters	SKY 525	360	No
Terra	ML 200	100	No
Terra	ML 360	360	No
Terra	ML 720	720	No

VHF NAVCOMMS

Maker	Model	Channels	Current (1980)
Bayside	BEI 880	100/ 90	No
Bendix	CNS 220B	360/360	No
Bendix	CNS 240	360/760	No

VHF NAVCOMMS

Maker	Model	Channels	Current (1980)
Bendix	M 450	100/360	No
Bendix	CN 2013A	200/720	Yes
Bendix	CN 2012A	200/720	Yes
Bendix	CN 2011A	200/720	Yes
Cessna ARC	317A/G	380/360	No
Cessna ARC	513	190/ 90	No
Cessna ARC	514A/R	80/ 90	No
Cessna ARC	515R	100/360	No
Cessna ARC	516A	80/ 90	No
Cessna ARC	522A	200/360	No
Cessna ARC	RT 308C	160/360	No
Cessna ARC	RT 328C	200/360	No
Cessna ARC	RT 328D	200/720	No
Cessna ARC	RT 328T	200/720	No
Cessna ARC	RT 385A	200/720	Yes
Cessna ARC	RT 422A	200/360	No
Cessna ARC	RT 428A	200/720	No
Cessna ARC	RF 485A	200/720	Yes
Cessna ARC	RT 508A	100/100	No
Cessna ARC	RT 517R	100/ 90	No
Cessna ARC	RT 522A	200/360	No
Cessna ARC	RT 528A/E	200/360	No
Cessna ARC	RT 540A	100/360	No
Edo-Aire	RT 553	200/360	Yes
Edo-Aire	RT 553A	200/720	Yes
Edo-Aire	RT 551/2	200/360	Yes
Edo-Aire	RT 551A/2	200/720	Yes
Edo-Aire	RT 661/2	200/360	Yes
Edo-Aire	RT 563	200/360	Yes
Edo-Aire	RT 661A/2	200/720	Yes
Genave	Alpha/190	100/ 90	No
Genave	Alpha/200	100/100	Yes
Genave	Alpha/300	100/360	No
Genave	Alpha/360	100/360	No
Genave	Alpha/500	200/360	No
Genave	Alpha/600	200/360	No
Genave	GA 1000	200/720	Yes
King	KX 100A	190/ 90	No
King	KX 120	100/360	No

<h1 style="text-align:center">VHF NAVCOMMS</h1>

Maker	Model		Current (1980)
King	KX 130	100/360	No
King	KX 145	200/720	Yes
King	KX 150	100/100	No
King	KX 160	100/360	No
King	KX 170B	200/720	Yes
King	KX 175B	200/720	Yes
Mentor	M 135	100/360	No
Mentor	M 400	100/360	No
Mentor	M 450	100/360	No
Narco	Mark 2	Tune/27	No
Narco	Mark 3	190/ 90	No
Narco	Mark 10	190/360	No
Narco	Mark 12	100/ 90	No
Narco	Mark 12	100/360	No
Narco	Mark 12A	100/ 90	No
Narco	Mark 12A	100/360	No
Narco	Mark 16	200/360	No
Narco	Mark 24	100/360	No
Narco	Escort 110	100/110	No
Radair	300	200/360	No
Radair	R 300	200/360	No
Skycrafters	605	100/360	No
Skycrafters	SA 1036	100/360	No
Terra	ML 200	200/100	No

<h1 style="text-align:center">VHF NAVS</h1>

Maker	Model		Current (1980)
Bayside	BEI 800	100	No
Bendix	RN 242A	200	No
Cessna ARC	R 442A	200	No
Cessna ARC	525A	100	No
Collins	51R7	380	No
Collins	VIR 350	400	Yes
Collins	VIR 351	200	Yes
Edo-Aire	R 552	200	Yes
Edo-Aire	R 554	200	Yes
Edo-Aire	R 662	200	Yes

Maker	Model		Current (1980)
Edo-Aire	R 664	200	Yes
King	KN 53	200	Yes
King	KNS 80	200	Yes
King	KNS 81	200	Yes
Mentor	M 200	200	Yes
Narco	Mark 8	190	No
Narco	NAV 10	200	No
Narco	NAV 11	200	No
Narco	NAV 12	200	No
Narco	NAV 14	360	No
Narco	NAV 112	200	No
Narco	NAV 114	360	No
Narco	NAV 121	200	Yes
Narco	NAV 122	200	Yes
Narco	NAC 124	200	Yes
Radair	240	200	No
Terra	ML 105	100	No
Terra	R 200	200	Yes

DMES

Maker	Model	Current (1980)
Bendix	DME 2030	Yes
Cessna ARC	RTA 476A	Yes
Cessna ARC	RTA 876A	Yes
Collins	DME 40	No
Collins	DME 451	Yes
Edo-Aire	RT 888	Yes
King	KN 60C	No
King	KN 61	No
King	KN 62A	Yes
King	KN 63	Yes
King	KN 65	No
King	KNS 80	Yes
Narco	UDI 2	No
Narco	UDI 3	No
Narco	UDI 4	No
Narco	DME 70	No
Narco	DME 190	Yes
Narco	DME 195	Yes
RCA	AVQ 75	No

Maker	Model	Current (1980)
Bendix	DF 2071A	Yes
Bendix	T 12C	No
Bendix	T 12D	No
Cessna ARC	21A	No
Cessna ARC	318A	No
Cessna ARC	324A	No
Cessna ARC	346	No
Cessna ARC	R 446A	Yes
Cessna ARC	R 546E	Yes
Cessna ARC	R 846A	No
Collins	ADF 650A	Yes
Edo-Aire	R 556	No
Edo-Aire	R 556D	No
Edo-Aire	R 556E	Yes
Genave	Sigma 1500	Yes
Kett	Polaris	No
Kett	Polaris X	No
Kett	Polaris XX	No
King	KR 80	No
King	KR 85	No
King	KR 86	Yes
King	KR 87	Yes
Motorola	ADF T 12B	No
Narco	ADF 29	No
Narco	ADF 31	No
Narco	PDF 35	No
Narco	ADF 140	No
Narco	ADF 141	Yes

Transponders

Maker	Model	Current (1980)
Bendix	TR 2061A	Yes
Cessna ARC	RT 359A	Yes
Cessna ARC	RT 459A	Yes
Collins	TDR 950	Yes
Collins	TDR 950L	Yes
Edo-Aire	RT 667	Yes
Edo-Aire	RT 777	Yes

Maker	Model	Current (1980)
Edo-Aire	RT 887	Yes
Genave	Beta/5000	Yes
King	KT 76A	Yes
King	KT 78A	Yes
Narco	AT 6A	No
Narco	AT 50A	No
Narco	AT 150	Yes
Narco	UAT 1	No
Regency	505	No
Terra	R 250	Yes
Wilcox	814B	No

Audio Control Panels

Maker	Model	Current (1980)
Bendix	AS 248A	No
Cessna ARC	F 1010B	Yes
Collins	AUD 250	Yes
Collins	AUD 251H	Yes
Collins	AMR 350	Yes
Edo-Aire	A 550	Yes
Edo-Aire	AM 550	Yes
Edo-Aire	AM 660	Yes
Genave	TAU 200	Yes
King	KA 37	No
King	KA 134	Yes
King	KMA 20	No
King	KMA 24	Yes
Narco	CP 125	No
Narco	CP 126	No
Narco	CP 127	No
Narco	CP 135	Yes
Narco	CP 136	Yes
Narco	CP 136T	Yes

RNAVS

Maker	Model	Current (1980)
Bendix	NP 2041A	Yes
Cessna ARC	RN 478A	Yes

Maker	Model	Current (1980)
Collins	ANS 351	Yes
Foster	RNAV 511	Yes
Foster	RNAV 611	Yes
Foster	RNAV 612	Yes
King	KN 74	No
King	KNS 80	Yes
King	KNS 81	Yes
King	KNC 610	Yes
Narco	CLC 60	No
Narco	CLC 60A	No
Narco	RNAV 161	No

Glideslopes

Maker	Model	Current (1980)
Bendix	204A	No
Bendix	GM 247A	No
Cessna ARC	R 31A	No
Cessna ARC	R 443B	Yes
Cessna ARC	R 543B	No
Collins	51 V5	No
Collins	GLS 350	Yes
Edo-Aire	CID 774	No
Genave	Phi/20	No
Genave	Phi/40	No
King	KGS 680	No
King	KGM 690	No
King	KN 70	No
King	KN 73	No
King	KN 75	Yes
King	KNS 80	Yes
King	KNS 81	Yes
Narco	UGR 1	No
Narco	UGR 2	No
Narco	UHR 2A	Yes
Regency	303	No
Terra	R 240	Yes
Wilcox	800 A, B	No

Index